Trần, Nguyễn Hoài An

Melt Spinning and Characterization of Biodegradable Micro- and Nanofibrillar Structures from Poly(lactic acid) and Poly(vinyl alcohol) Blends

TUDpress
2016

Bibliografische Information der Deutschen Nationalbibliothek
Die Deutsche Nationalbibliothek verzeichnet diese Publikation in der
Deutschen Nationalbibliografie; detaillierte bibliografische Daten sind
im Internet über http://dnb.d-nb.de abrufbar.

Bibliographic information published by the Deutsche Nationalbibliothek
The Deutsche Nationalbibliothek lists this publication in the Deutsche
Nationalbibliografie; detailed bibliographic data are available in the
Internet at http://dnb.d-nb.de.

ISBN 978-3-95908-051-4

© 2016 w.e.b. Universitätsverlag & Buchhandel
Eckhard Richter & Co. OHG
Bergstr. 70 | D-01069 Dresden
Tel.: 0351/47 96 97 20 | Fax: 0351/47 96 08 19
http://www.tudpress.de

TUDpress ist ein Imprint von w. e. b.
Alle Rechte vorbehalten. All rights reserved.
Gesetzt vom Autor.
Printed in Germany.

Melt spinning and characterization of biodegradable micro- and nanofibrillar structures from poly(lactic acid) and poly(vinyl alcohol) blends

(Schmelzspinnen und Charakterisierung von biologisch abbaubaren mikro- und nanofibrillären Strukturen aus Mischungen der Polymilchsäure und Polyvinylalkohol)

Der Fakultät Maschinenwesen

der

Technischen Universität Dresden

zur

Erlangung des akademischen Grades

Doktoringenieur (Dr.-Ing.)

angenommene Dissertation

M. Sc. Tran, Nguyen Hoai An

geboren am: 20.08.1980   in: Đồng Nai, Việt Nam

Tag der Einreichung: 03.11.2015

Tag der Verteidigung: 22.03.2016

Promotionskommission:

Vorsitzender:	Herr Prof. Dr.rer.medic. Hans P. Wiesmann
Gutachter:	Herr Prof. Dr.rer.nat.habil. Gert Heinrich
	Herr Prof. Dr.rer.nat.habil. Dirk W. Schubert
Mitglieder:	Herr Prof. Dr.-Ing.habil. Dipl.-Wirt. Ing. Chokri Cherif
	Herr Prof. Dr.rer.nat.habil. Thomas Bley

To My Dearest Mother,
My Beloved Wife Phuong,
and Daughters Nghi and Dung

Vorwort

Die vorliegende Arbeit entstand während meiner Tätigkeit als wissenschaftlicher Mitarbeiter im Leibniz-Institut für Polymerforschung Dresden e. V. (IPF Dresden e.v.), Teilinstitut für Polymerwerkstoffe, Abteilung Verarbeitungsprozesse, Forschungsgruppe Fadenbildung im Zeitraum von Oktober 2011 bis September 2015.

Mein besonderer Dank gilt Herrn Prof. Dr. rer. nat. habil. Gert Heinrich für die Übernahme der Betreuung und die finanzielle Unterstützung. Großer Dank gilt Herrn Prof. Dr.-Ing. Udo Wagenknecht für die organisatorische Unterstützung. Herrn Prof. Dr. rer. nat. habil. Dirk W. Schubert danke ich für sein Interesse an meiner Arbeit sowie die Übernahme des Koreferates.

Weiterhin danke ich insbesondere Herrn Dr. rer. nat. Harald Brünig für sein großes wissenschaftliches Interesse, die fundierte fachliche Betreuung, vielfache angeregte Diskussionen sowie für die sorgfältige Durchsicht der Manuskripte. Frau Dr.-Ing. Claudia Hinüber und Herrn Dipl.-Ing. Pit Wiedermann danke ich für die hilfreichen Diskussionen und die guten Ergebnisse.

Vielen Dank an Frau Maria Auf der Landwehr für die gute Zusammenarbeit sowie ihre Geduld und unermüdliche Hilfe bei der Durchführung der SEM-Aufnahmen. Herrn Mathias Häschel und Herrn Norbert Smolka danke ich vielmals für die technische Unterstützung und freundliche Hilfsbereitschaft bei den Spinnprozessen. Herrn Mathias Pech, Herrn Mathias Ulrich sowie den Kollegen der Werkstatt danke ich für die Konstruktion und Herstellung der Versuchsaufbauten.

Herrn Dr.-Ing. Roland Vogel danke ich für die hervorragende fachliche Unterstützung sowie seine Bereitschaft mir jede Frage hinsichtlich der rheologischen Untersuchungen zu beantworten. Herrn Dr. Mikhail Malanin und Herrn Dr. Andreas Leuteritz danke ich für ihre immer freundliche Hilfestellung und fachliche Betreuung bei den FTIR-Messungen. Frau Dipl.-Chem. Liane Häußler danke ich ganz herzlich für die Durchsicht der DSC-Ergebnisse

Vorwort

und die wertvollen Hinweise. Herrn Dr. Jürgen Pionteck danke ich sehr für freundliche Gespräche und für die fachliche Unterstützung zur Berechnung der Grenzflächenspannung.

Den Kolleginnen, Kollegen und Mitarbeitern des IPF Dresden e. V. danke ich für die nette Zusammenarbeit. Insbesondere bei Frau Anna Ivanov, Frau Dipl.-Lab.-Chem. Regine Boldt, Frau Ivonne Hasselhorst, Frau Kristina Eichhorn, Herrn Lutz Peitzsch, Herrn Andreas Janke und Herrn Holger Scheibner danke ich für die Geräteeinweisung bzw. die Durchführung meiner umfangreichen experimentellen Untersuchungen für MFI, SEM, AFM sowie am Lichtmikroskop und für Zugversuche.

Herrn Prof. Dr. Stoyko Fakirov bedanke ich mich für seine wissenschaftlichen Ideen und sein Engagement zum Thema.

Für den wissenschaftlichen Gedankenaustauch in freundschaftlich entspannter Atmosphäre in den Doktorandenseminaren sowie bei Tagungen bedanke ich mich bei allen IPF Doktoranden, insbesondere bei Guido Sommer, Debdipta Basu, Dmytro Ivaneiko und Marco Liebscher.

Der vietnamesischen Regierung und dem Ministerium für die Bildung und Ausbildung danke ich für die finanzielle Unterstützung im Rahmen des Promotionsstipendiums. Nicht zuletzt gebührt der Deutschen Forschungsgemeinschaft für die finanzielle Förderung im Rahmen des Forschungsprojekts „Entwicklung eines neuartigen Filamentgarnes" (BR 1886/6-1) Dank.

Trần, Nguyễn Hoài An					Dresden, im Oktober 2015

Eidesstattliche Erklärung

Hiermit versichere ich, dass ich die vorliegende Arbeit ohne unzulässige Hilfe Dritter und ohne Benutzung anderer als der angegebenen Hilfsmittel angefertigt habe; die aus fremden Quellen direkt oder indirekt übernommenen Gedanken sind als solche kenntlich gemacht.

Dresden, den 22.10.2015

Tran, Nguyen Hoai An

Abstract

The present study mainly focuses its attention on the investigation and the development of the biodegradable and biocompatible nanofibrillar poly(lactic acid) structures from poly(lactic acid) (PLA) and poly(vinyl alcohol) (PVAL) blends using the conventional melt spinning process (CMS) of thermoplastic polymer fibers. The CMS itself is a high productivity, solvent free and low energy consumption process. Owing to the combination of these outstanding benefits of the CMS and the "green" properties of PLA and PVAL, the studied and developed process provides an optimal solution for dealing with relevant challenges of our present age: the economic, highly productive and extremely eco-friendly fabrication of highly specialized products for biomedical applications such as scaffolds for tissue engineering. In that regard, samples of PLA, PVAL and their blends are extruded in a co-rotating twin-screw microextruder. The blended extrudates and filaments are then characterized with respect to their micro-/nanofibrillar morphology as well as thermal and rheological properties to fix the PLA/PVAL blend ratio and find the appropriate melting temperature profiles for the melt spinning process of PLA/PVAL blends. Considering the effect of take-up velocity and off-line draw ratio on the textile-physical properties of PLA/PVAL blend filaments as well as the dispersed PLA phase morphology after removal of the water soluble PVAL matrix, an optimum set of spinning conditions for the highly reproducible and stable melt spinning process of PLA/PVAL blends is proposed.

Another research activity gives special attention on the study of the morphological evolution of PLA/PVAL blend filaments along the spinline. Emphasis of the investigation is given on considering the axial strain rate and tensile stress combined to other filament parameters such as temperature, diameter, and apparent elongational viscosity that influence the final state of deformation of dispersed PLA phase in PLA/PVAL blends. Owing to a new special self-constructed fiber-capturing device, pieces of the PLA/PVAL blend filaments are collected at different locations along the spinline. The morphology of the PLA/PVAL captured blend

Abstract

filaments in both longitudinal and cross-sectional directions are then investigated using scanning electron microscopy (SEM) technique. The axial strain rate (ASR) at different zones along the spinline is calculated from velocity data using Laser Doppler Velocimetry (LDV) technique. The tensile stress is determined from the force balance equation. The filament temperature profile along the spinline is obtained both by online measurement using an infrared camera and by simulation using the energy balance equation of the thin filament model approximation. Comparing all the spinline parameter profiles, especially the ASR and tensile stress, with the morphological evolution of dispersed PLA phase along the spinline, the mechanism of the fibrillation process within fiber formation zone is elucidated. By adjusting the melt spinning conditions, for instance take-up velocity and mass flow rate, the final morphology of dispersed PLA phase in PLA/PVAL blends is controlled. It is found that the increasing velocity and tensile stress in the fiber formation zone are mainly responsible for stretching and coalescence of the dispersed PLA phase into nanofibrils in the PVAL matrix of polymer blends.

In the present study, the influence of spinning conditions on the fibrillation process of polymer blends, on the example of PLA/PVAL blend systems, in an elongational flow within fiber formation zone are systematically and thoroughly investigated for the first time. The findings of the current work provide valuable insight into the morphology development of dispersed PLA phase and present basic requirements for producing nanofibrillar structures using a conventional melt spinning process.

Kurzfassung

In der vorliegenden Arbeit werden die Herstellung und Entwicklung von neuartigen, biologisch abbaubaren und biokompatiblen nanofibrillären Faserstrukturen aus Mischungen der thermoplastischen Polymere Polymilchsäure (Polylactic acid, PLA) und Polyvinylalkohol (PVAL) mit Hilfe des konventionellen Schmelzspinnprozesses untersucht. Das energetisch günstige Schmelzspinnverfahren ist hochproduktiv und frei von Lösungsmitteln. In der Kombination dieser Vorteile mit den „grünen" Eigenschaften der verwendeten PLA und PVAL, bietet das untersuchte Verfahren eine mögliche Lösung für eine wichtige Herausforderung unserer Zeit: die ökonomisch effektive und gleichzeitig umweltgerechte Herstellung hochspezialisierter Produkte, wie z. B. von Scaffolds im Bereich des medizinischen Tissue Engineering. Zu diesem Zweck wurden die Polymere PLA und PVAL in verschiedenen Verhältnissen im Zweischnecken-Mikroextruder compoundiert. Die Morphologie sowie die thermischen und rheologischen Eigenschaften der Blends und der daraus hergestellten Filamente wurden bestimmt, um ein optimales Mischungsverhältnis und ein geeignetes Temperaturprofil im Schmelzspinnprozess der PLA/PVAL-Blends einzustellen. Des Weiteren wurden der Einfluss der Abzugsgeschwindigkeit und des offline-Verstreckverhältnisses auf die textilphysikalischen Eigenschaften der PLA/PVAL-Blendfilamente ebenso wie auf die Morphologie der PLA-Phase (nach Entfernung der wasserlöslichen PVAL-Matrix) untersucht. Im Ergebnis werden für den reproduzierbaren und stabilen Schmelzspinnprozess der PLA/PVAL-Blends optimale Spinnbedingungen vorgeschlagen.

Einen weiteren Schwerpunkt der Arbeit stellen die Untersuchungen zur Herausbildung der morphologischen Struktur der PLA/PVAL-Blendfilamente entlang der Spinnlinie dar. Insbesondere betrachtet werden dabei der Einfluss der axialen Dehnungsrate und der Fadenzugspannung im Zusammenhang mit anderen Filamentparametern, wie Filamenttemperatur, -durchmesser und (scheinbare) Dehnviskosität, die den Endzustand der

Kurzfassung

Deformation der dispersen PLA-Phase im PLA/PVAL-Blend beeinflussen. Mit Hilfe einer speziell hierfür entwickelten Faser-Klemm-Vorrichtung wurden dazu kurze Faserstücken an verschiedenen Stellen entlang der Spinnlinie entnommen. Deren Morphologie wurde sowohl im Längs- als auch im Querschnitt mit Hilfe der Rasterelektronenmikroskopie (REM) untersucht. Die axiale Dehnungsrate an den entsprechenden Positionen der Spinnlinie wurde aus den mittels Laser-Doppler-Anemometrie gewonnenen Geschwindigkeitsdaten berechnet. Die Fadenspannung wurde mittels Tensiometer unter Berücksichtigung der Beziehungen für das Kräftegleichgewicht am Filament bestimmt. Der Temperaturverlauf in der Fadenbildungszone wurde sowohl durch online-Messungen mittels Infrarot-Thermographie als auch durch Anwendung der Energiebilanzgleichung im Rahmen des eindimensionalen stationären Fadenbildungsmodells (thin filament model approximation) bestimmt. Durch den Vergleich und die Analyse der Beziehungen zwischen den Filamentparametern und zum Verlauf der Entwicklung der Morphologie der PLA-Phase entlang der Spinnlinie kann der Mechanismus des Fibrillierungsprozesses in der Fadenbildungszone aufgeklärt werden. Veränderungen der Spinnbedingungen, wie beispielsweise der Abzugsgeschwindigkeit oder des Durchsatzes, verändern die Morphologie der dispersen PLA-Phase. Es wird deutlich, dass vor allem zunehmende Geschwindigkeit und Spannung in der Fadenbildungszone verantwortlich sind für die Deformation und Koaleszenz der dispersen PLA-Phase und für die Entstehung der PLA-Nanofibrillen innerhalb der PVAL-Matrix.

In der vorliegenden Arbeit wird der Einfluss der Spinnbedingungen auf die Herausbildung von Fibrillen in Polymerblends innerhalb der Fadenbildungszone des Schmelzspinnprozesses am Beispiel eines PLA/PVAL-Systems erstmalig systematisch und umfassend untersucht. Die Erkenntnisse der Arbeit führen zu einem besseren Verständnis bezüglich der Entwicklung der Morphologie der PLA-Phase und stellen grundlegende Voraussetzungen für die Herstellung nanofibrillärer Faserstrukturen mit Hilfe eines konventionellen Schmelzspinnverfahrens dar.

Contents

Vorwort iii

Abstract vi

Kurzfassung viii

Contents x

List of symbols and abbreviations xiii

1 Introduction 1

2 Background and literature survey 7

    2.1 Materials 7

        2.1.1 Introduction to materials 7

        2.1.2 Poly(vinyl alcohol) (PVAL) 8

        2.1.3 Poly(lactic acid) (PLA) 11

    2.2 Polymer blends with micro-nanofibrillar phase morphology 12

        2.2.1 Polymer-polymer miscibility and phase separation 12

        2.2.2 Phase morphologies of immiscible polymer blends 13

        2.2.3 The formation of micro-nanofibrillar morphology in an elongational flow 14

        2.2.4 Theoretical considerations of droplet deformation and break-up 15

    2.3 Theoretical considerations of thin filament model in the melt spinning 19

        2.3.1 Mass balance equation 20

        2.3.2 Force balance equation 21

Contents

  2.3.3 Energy balance equation 23

3 Experimental 26

 3.1 Materials 26

  3.1.1 Poly(vinyl alcohol) (PVAL) 26

  3.1.2 Poly(lactic acid) (PLA) 28

 3.2 Processing 29

  3.2.1 Melt mixing 29

  3.2.2 Melt spinning 30

  3.2.3 Off-line drawing process 32

 3.3 Characterization and on-line measurements 33

  3.3.1 Attenuated total reflection-Fourier transform infrared spectroscopy 33

  3.3.2 Differential scanning calorimetry (DSC) 34

  3.3.3 Thermogravimetric analysis (TGA) 34

  3.3.4 Rheological measurements 34

  3.3.5 On-line filament speed measurement 35

  3.3.6 On-line filament temperature measurement 37

  3.3.7 On-line tension measurement 40

  3.3.8 Fiber-capturing device 40

  3.3.9 Filament diameter measurement via light microscopy 41

  3.3.10 Mechanical characterization 42

  3.3.11 Morphology characterization 44

4 Results and discussion 47

 4.1 Miscibility of PLA/PVAL blends 47

  4.1.1 Glass transition temperature 48

  4.1.2 ATR-FTIR spectroscopy 50

  4.1.3 SEM and atomic force microscopy (AFM) images 54

 4.2 Melt Spinning of nanofibrillar structures from PLA/PVAL blends 57

4.2.1	Blend ratio and spinning temperature profiles for the melt spinning process	57
4.2.2	Textile-physical properties of PLA/PVAL 30/70 filaments	65
4.2.3	Morphology of nanofibrillar PLA structures from PLA/PVAL filaments	68
4.2.4	Purity of nanofibrillar PLA structures (PLA scaffolds)	73
4.2.5	Mechanical properties of nanofibrillar PLA structures (PLA scaffolds)	75
4.2.6	Discussion	76
4.3	Characterization of PLA/PVAL monofilament profiles	77
4.3.1	Filament temperature profiles	77
4.3.2	Velocity and velocity gradient along the spinline	80
4.3.3	Filament diameter profiles	82
4.3.4	Tensile force, tensile stress, and apparent elongational viscosity profiles	85
4.3.5	Discussion	89
4.4	Morphology development of PLA/PVAL blends in an elongational flow	93
4.4.1	Morphology development of PLA/PVAL 30/70 blends along the spinline	93
4.4.2	Controlling the micro- and nanofibrillar PLA morphology	102
5	Conclusions and future works	132
5.1	Conclusions	132
5.2	Future works	135
References		136
List of figures		148
List of tables		157
Appendix A – D		A – D

List of symbols and abbreviations

All symbols and abbreviations are defined at their first appearance. If they used only one or two times in the dissertation, they are not listed here.

Symbols

a, b, c	coefficient constant	–
$A(x)$	cross-sectional area of filament	μm^2
B	droplet diameter in affine deformation theory	μm
Ca	capillary number	–
Ca_c	critical capillary number	–
Ca^*	reduced capillary number	–
c_f	air friction coefficient	–
$d_{CED}, \bar{d}_{CED}$	circular equivalent diameter (CED), mean CED of dispersed PLA phase in cross-sectional PLA/PVAL blend extrudates after etching PLA phase	μm
$d, \bar{d}$	diameter and mean diameter of dispersed PLA phase from PLA/PVAL blend extrudates after removing PVAL matrix	μm
D	droplet deformation	–
D_0	diameter of filament at the die exit	μm
$D(x)$	diameter of filament at a distance x along the spinline	μm
E_0	Young's modulus	$cN \cdot tex^{-1}$
D_L	take-up filament diameter	μm
$F_{aero}, F_{grav}, F_{inert}, F_{surf}$	air drag, gravitational, inertial, surface tension force	cN
F_0	initial force	cN

List of symbols and abbreviations

F_L	take-up force	cN
Q	mass flow rate	$g \cdot min^{-1}$
L_S	solidification length	cm
Nu	Nusselt number	–
Re	Reynolds number	–
t	time	s
t_S	solidification time	s
$T_{g,PLA}, T_{g,PVAL}$	glass transition temperature of PLA, PVAL	°C
$T_{m,PLA}, T_{m,PVAL}$	melting temperature of PLA, PVAL	°C
$T(x)$	filament temperature at distance x along the spinline	°C
$v(x)$	filament velocity at distance x along the spinline	$m \cdot min^{-1}$
v	take-up velocity	$m \cdot min^{-1}$
$\dot{V}$	the volumetric flow rate	$cm^3 \cdot min^{-1}$
x	distance from the spinneret along the spinline	cm
x_S	distance at solidification point	cm

Greek symbols

η	viscosity	$Pa \cdot s$
η_d	viscosity of the dispersed phase	$Pa \cdot s$
η_m	viscosity of the matrix	$Pa \cdot s$
η_{app}	apparent elongational viscosity	$Pa \cdot s$
η_{air}	dynamic viscosity of air	$Pa \cdot s$
$\eta^*(\omega)$	complex viscosity	$Pa \cdot s$
ε_R	elongation at break	%
$\dot{\varepsilon}$	elongation rate	s^{-1}
ϱ_{air}	mass density of air	$kg \cdot m^{-3}$
ϱ_P	mass density of polymer	$g \cdot cm^{-3}$
$\dot{\gamma}$	shear rate / rate of deformation tensor	s^{-1}
λ	viscosity ratio	–
σ	tensile stress	Pa
Γ	interfacial stress	$mN \cdot m^{-1}$

Abbreviations

AFM	atomic force microscopy
APEV	apparent elongational viscosity
ASR	axial strain rate
ATR-FTIR	attenuated total reflection-Fourier transform infrared
CED	circular equivalent diameter
DR	draw ratio
DSC	differential scanning calorimetry
MFC	microfibrillar composite
MFF	matrix-fibrillar fiber
MFI	melt flow index
NFC	nanofibrillar composite
P	position
PLA	poly(lactic acid)/poly lactide
PVAL	poly(vinyl alcohol)
SCD	spinning condition
SEM	scanning electron microscopy

Chapter 1
Introduction

Biocompatible and biodegradable nanofibers came recently into the focus of considerable academic and industrial interest due to their unique features (e.g. large surface-area-to-volume ratio, extremely small pore dimensions) and their great potential not only for tissue engineering but also for various biomedical applications including wound dressings, membrane filters, and drug delivery systems [1, 2]. Biocompatible and biodegradable nanofibers refer to fibers with diameter of 100 nm or less[1] [3] fabricated from biocompatible and biodegradable polymeric materials.

Several fabrication processes have been developed to produce nanofibers. They can be divided into three types [4-6]: self-ordering (e.g. phase separation, template synthesis), random (electrospinning, melt blown, centrifugal spinning, etc.), and filament type process including conjugated spinning (islands-in-the-sea or sea-island-type) and polymer blend melt spinning. The both processes of the filament type are similar to the conventional melt spinning method. Therefore, they have many advantages of a conventional melt spinning process, which is high productive and environmentally-friendly due to the absence of solvents. Furthermore, the production form "filament type" of these processes shows high flexibility that can be subjected to textile processing in different textile processes such as weaving, knitting, and embroidery before generating the micro-/nanofibers.

Sea-island-type conjugated spinning is the melt spinning method of manufacture for bicomponent or conjugate fibers. These fibers, in which many island fibers are arranged in a sea component, are formed by combining two separate polymer streams just before their delivery to islands-in-the-sea dies [7, 8]. The micro-/nanofibers are then obtained by the extraction of the sea component. Even using this melt spinning method, it is difficult to

---

[1] In textile industry, this definition is often extended to include fibers having diameters smaller than 500 nm

Introduction

produce island fibers having a diameter less than 100 nm. Nakata et al. [5] reported that PET nanofibers with a diameter of 39 nm from PET/nylon-6 island/sea filament by using sea-island-type conjugated melt spinning with further laser-heated flow drawing and removal of the sea component. To dissolve the nylon-6 component from PET/nylon-6 blend filaments, formic acid and ultrasonic waves were used. We note that this dissolvent process is considered to be not environmental friendly. The conjugated melt spinning method is also expensive because it involves the extrusion of bicomponent fibers, which can be made by using a special desired spinneret-hole configuration with separate delivery of two polymer melts. This involves additional capital costs for the second polymer melt system and the special spinneret packs. Moreover, cleaning, maintaining the parts, and inspection of such spinnerets are more complex and expensive compared to monocomponent spinning [9, 10].

Unlike the sea-island-type conjugated melt spinning, the melt spinning of polymer blend is relative simple, at least in principle, because it can be similar to the melt spinning process used for making single polymer fibers [7, 11]. The melt spinning of polymer blend can be processed in two ways: (1) Two incompatible polymers can be dry-blended in granule form, intimately mixed in a twin screw-extruder on the melt spinning machine, and melt spun into blend fibers, the so-called matrix-fibrillar fibers (MFFs)[1] or "fibrils in a matrix" (Figure 1.1a and 1.1b) [7, 12]; (2) The granules of two polymers can be mixed in any conventional twin-screw extruder and extruded through a single orifice die forming a blend extrudate product. Blend extrudates are then pelletized before being reextruded through a single-screw extruder and melt spun in to MFFs on the conventional melt spinning machine. The former one is more often used in the melt spinning process to prevent heating effects of the preblend processes. Like the nanofibers produced using the conjugated melt spinning method, the nanofibers in the melt spinning of polymer blend are then obtained by the extraction of the matrix from MFFs (Figure 1.1c).

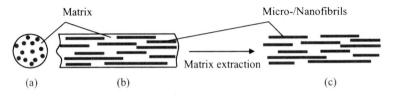

Figure 1.1 Schematic representation of cross- and longitudinal section of blend fibers/MFFs (a, b), and micro-/nanofibrils (c)

---

[1] The term "matrix-fibrillar fibers" are not considered as bicomponent fibers in the USA where they are known as matrix fibers [12]

Introduction

The complexity of the manufacturing of the nanofibers using the melt spinning of polymer blend is known by the choice of suitable polymer components and the control over the melt spinning parameters. It is well established that the formation of micro- and nanofibrillar phase morphology of two immiscible polymer blends are substantially affected by the properties of the blend components (e.g. the blend ratios, viscoelasticity of components, and interfacial tension between components) as well as the processing conditions (temperature, time, draw ratio, shear rate, shear stress, total strain, etc.) [13-19].

The melt spinning process of polymer blends that results into micro-/nanofibers has been known since 1960's and the first patent of the fibrillation of polymer blends in elongational flow was issued to Miller and Merriam in 1963 [20]. However, polymeric micro-/nanofibers obtained by melt extrusion and melt spinning of polymer blends with further extraction of the matrix polymer have become again the subject of intensive investigations over the last five to ten years due to a tremendous increase for economic and environmental concerns. So far, typical polymer blends (dispersed/matrix) used included polyethylene (PE)/polystyrene (PS) [21], PE/poly(vinyl acetate) (PVA), PE/polyamide 6 (PA6), PA6/polypropylene (PP) [22-24], poly(butylene terephthalate) (PBT)/PP [25, 26], poly(ethylene terephthalate) (PET)/polyamide 66 (PA66), PET/PE [27], PET/PP [28], and various thermoplastic polymers (TPs)/cellulose acetate butyrate (CAB) [29].

Almost all above mentioned investigations [20-29] used the various solvents such as toluene, formic acid, decalin (decahydronaphthalene), hot xylene, and acetone to remove PS, PA6/PA66, PP, PE, and CAB, respectively. Although some solvents are less hazardous than others, all solvents can cause toxic effects on the nervous system, respiratory system, skin, eyes, and internal organs to some degree[1]. Furthermore, all dissolvent processes seem to be not environmentally-friendly and economical. To solve such problems, Robeson et al. [30] and Tsebrenko et al. [31] used thermoplastic poly(vinyl alcohol) (PVAL) as a matrix component to produce the PP microfibers from PP/PVAL blends with the extraction of PVAL using water only. Recently, Fakirov, Bhattacharyya, Lin et al. [32-34] have employed also PVAL as a matrix component and poly(lactic acid) (PLA) as a dispersed phase to fabricate the micro- and nanofibrillar PLA scaffolds by using melt extrusion through capillary die followed by drawing with further extraction of PVAL matrix. The PLA scaffolds were prepared in this way by excluding any contact with organic solvents or toxic components. The results of the biomedical testing with living cells showed that PLA scaffolds are promising candidates for

---

[1] Source: UIC Health and Safety in the Arts Library

Introduction

tissue engineering or as carriers for controlled drug delivery. These results open many opportunities for the fabrication of micro-/nanofibers for biomedical applications. However, these preliminary investigations by Fakirov et al. [32-34] have been limited in melt extrusion process with the absence of a strong elongational flow like that exists in the melt spinning process. It is also reported that the microfibrils obtained having a diameter and length of about 1 µm and 20 µm length, respectively. It is of interest to note that the melt spinning process with the presence of a combination of shear and elongational flow fields seem to be more effective in producing finer and more continuous fibrils.

In that regard, the main objective of this work is to produce continuous biodegradable and biocompatible fibers in nano-scale by using the melt spinning process of polymer blends and by selecting PVAL with the properties of "green polymer" as a matrix component. This fabrication process therefore provides an optimal solution to handle important scientific challenges at the present time: the economic, highly productive and extremely environmentally-friendly fabrication of highly specialized products for medical purposes as scaffolds for tissue engineering.

Recently, biodegradable nanofibrillar PLA structures with average diameter of 60 nm were successfully fabricated by melt spinning of PVAL and PLA blends with both single-screw [35] and twin-screw extruder [36] and subsequent removal of the PVAL matrix. However, many questions and problems regarding the melt spinning process of the PLA/PVAL blend for the formation of nanofibrillar structures still remain unanswered.

(1) Despite many efforts, the melt spinning of PVAL has not been succeeded industrially due to its thermal degradation and high viscosity [37-39]. Thermal and rheological behavior of PLA/PVAL blends is totally dominated by thermal and rheological properties of PVAL during melt spinning process. Therefore, the first goal of the present work is to find the melting temperature profiles and select an optimal set of spinning process parameters for the stable melt spinning process of PLA/PVAL blends on the melt spinning machine using twin-screw extruder.

(2) The formation of micro-and nanofibrillar phase morphology of polymer blends is also one of the most important questions of academic work during the last decades. Numerous studies on morphology development of polymer blends in twin-screw extruder as well as in the spinneret channel, and the effect of various parameters on it have been published [14, 40-54]. By contrast, few studies have tried to investigate the morphological variations of polymeric blend systems after extrusion from the spinneret

orifices into fibers [26, 55, 56]. However, these investigations were focused on the morphology of the dispersed phase in blend samples at only two positions along the spinline: Blend samples at the die exit and the as-spun fibers were taken from the bobbin. Therefore, an understanding of the morphology development of polymer blends and the mechanism of the fibrillation process within fiber formation zone in the melt spinning process is necessary to elucidate the nanofiber formation. This is the second goal of the present work.

(3) During melt spinning, the morphology development of polymer blends along the spinline is caused by the changes of filament parameters such as diameter, velocity, temperature, tensile force, tensile stress, and apparent elongational viscosity. The melt spinning of extremely high viscosity materials such PLA/PVAL blends at low take-up velocities up to 70 m·min^{-1} is very different from standard spinning velocities of 1000 – 4000 m·min^{-1} and high spinning velocities over 4000 m·min^{-1}. Therefore, the characterization of PLA/PVAL filament profiles in the melt spinning process is also the objective of the present work. It helps to explain the fibrillation process of the dispersed PLA phase in PLA/PVAL blend filaments along the spinline.

To fulfill these objectives, the following chapters of the dissertation will try to address concrete research questions/topics and solve problems. The thesis is organized as follows:

Chapter 2 gives a brief overview about the biodegradable polymers, especially poly(vinyl alcohol) (PVAL) and poly(lactic acid) (PLA) and their properties. The formation of polymer blends that results in micro- and nanofibrillar phase morphology is reviewed. This chapter highlights the theory of droplet deformation and break up of dispersed phase in the matrix component. Finally, the theoretical background of the melt spinning process including the thin filament model, heat and momentum transfer was also presented.

Chapter 3 describes the general features of the materials including thermal behaviours of PLA and PVAL pellets used in the present study. This chapter also focuses on the description of various experimental processes including the melt mixing, the melt spinning, and the hot drawing to produce the PLA/PVAL blend filaments. Finally, the chapter 3 is completed by descriptions of analytical tools and testing conditions used for the characterization of PLA/PVAL filament properties and PLA nanofibrillar structures obtained from PLA/PVAL filaments.

Introduction

Chapter 4 is divided into four main sections starting with the miscibility of PLA/PVAL blends, which is considered as basic requirement for producing the nanofibrillar structures. Next, a fabrication process for preparing continuous thermoplastic nanofibrillar PLA structures from PLA/PVAL 30/70 blend by using the conventional melt spinning method of thermoplastic polymer fibers is described. This melt spinning process was done on the industrial-scale melt spinning device at the Leibniz-Institut für Polymerforschung Dresden e. V. (IPF Dresden). In the second section 4.2, the thermal and rheological properties of the blend extrudates were analyzed to select an optimal set of spinning process parameters. The textile-physical properties of PLA/PVAL 30/70 blend filaments and off-line drawn blend filaments were also investigated. Furthermore, the morphology of dispersed PLA phase from both spun blend filaments and off-line drawn blend filaments after removing PVAL matrix were presented. In the third section 4.3, the filament parameter profiles such as filament temperature, filament velocity, filament diameter, tensile stress, etc. were investigated using online experimental measurements and the thin filament model of the melt spinning process. The last section 4.4 presents the morphological development of PLA/PVAL blends at different locations along the spinline under different spinning conditions. By considering the relationship between the changes in filament parameters in section 4.3 and the morphological development of PLA/PVAL blend filaments in section 4.4, the mechanisms of fibrillation process from rod like to nanofibrillar structures of dispersed PLA phase in a binary blend with PVAL matrix along the spinline was explained. It is worth noting that experimental results in the last two sections 4.3 and 4.4 are obtained from all the melt spinning experiments using the piston melt spinning device (small-scale melt spinning device) at the IPF Dresden.

Chapter 5 summarizes briefly the important contributions and final remarks of this thesis. Then, it identifies the promising applications of the PLA nanofibrillar structures and discusses the relevant and possible future works.

Chapter 2
Background and literature survey

2.1 Materials

2.1.1 Introduction to materials

Biocompatible and biodegradable polymeric materials are useful for various medical, agricultural, and packaging applications including wound dressings, membrane filters, and drug delivery systems [57, 58]. Many biodegradable polymers have been made to meet the needs of present times, even though the application of synthetic biodegradable polymer started in the latter of 1960s [59]. Current commercial biodegradable polymers are predominantly limited to aliphatic polyesters, polyethers, poly(vinyl alcohol), and polysaccharides [60-62].

As introduced in chapter 1, the purpose of the present work is to fabricate continuous thermoplastic nanofibrillar structures based on completely biodegradable systems by using the melt spinning method. The nanofibrillar structures are then obtained from polymer blend filaments after removing a matrix component by using an appropriate solvent. Normally, chemical solvents are used to remove the matrix component. Among the above mentioned biodegradable polymers, the water-soluble biodegradable poly(vinyl alcohol) (PVAL) is considered as the first candidate in blend systems, because PVAL is removed from polymer blend filaments using only distilled water. Based on a literature review of published articles and our empirical studies on the spinnability of polymer blends for the melt spinning process, poly(lactic acid) (PLA) is found as a potentially important component with PVAL. Furthermore, PVAL and PLA are the most widely fabricated biodegradable polymers, while other biodegradable polymers, such as poly(caprolacton) (PCL) and polyhydroxybutyrate (PHB), are produced in small quantities at the laboratory scale or at pilot plants [63]. The

Background and literature survey

following sections 2.1.2 and 2.1.3 present an overview of the properties and applications of these two biodegradable polymers: PLA and PVAL.

2.1.2 Poly(vinyl alcohol) (PVAL)

Poly(vinyl alcohol) (PVAL) was first discovered and prepared by Herrmann and Haehnel in 1924 by polymerizing vinyl acetate and then hydrolyzing poly(vinyl acetate) (PVAc) [64, 65]. It was first fabricated and commercially sold by the company Wacker (Germany) with the trademark Synthofil, and used as medical sutures [66, 67]. It was initially developed for fiber application by Japanese researchers Sakurada et al. in 1939 and was industrialized for the first time in the world by Kuraray in 1950 [67, 68]. It is also used for paper coating, warp sizing, adhesives (including protective colloids), and high-value added packaging films due to the antistatic and excellent barrier properties [69-71]. PVAL is now commonly used in biomedical and pharmaceutical fields due to its low protein adsorption properties, biodegradability, biocompatibility, high water solubility, and chemical resistance. Some of most common medical applications of PVAL are in soft contact lenses, eye drops, embolization particles, tissue adhesion barriers, and as artificial cartilage and meniscus [72-81].

PVAL is a carbon-carbon backbone polymer. Figure 2.1 shows the common chemical representations for PVAL. The physical, chemical, and mechanical properties of PVAL depend on their degree of polymerization, degree of hydrolysis, and distribution of the degree of hydrolysis [69, 82]. Table 2.1 summarizes the effect of the degree of hydrolysis and the average molecular weight of polymer on the physical properties of PVAL.

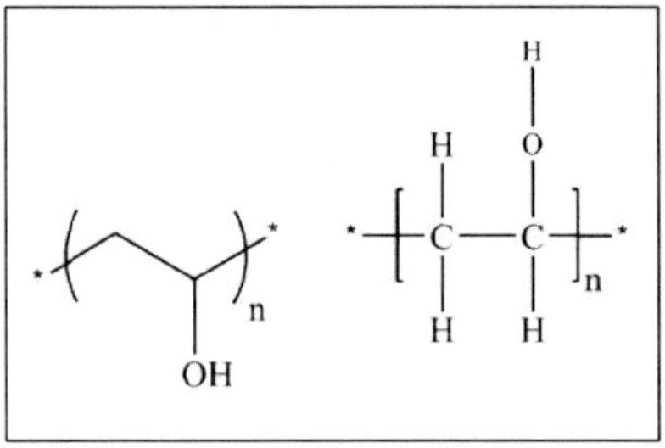

Figure 2.1 Chemical representations of PVAL

Table 2.1 Effect of hydrolysis degrees and molecular weight on the physical properties of PVAL, modified after [83]

Degrees of hydrolysis		Molecular weight	
Decreasing (↓)	Increasing (↑)	Decreasing (↓)	Increasing (↑)
- ↑ solubility	- ↑ viscosity	- ↑ solubility	- ↑ viscosity
- ↑ flexibility	- ↑ tensile strength	- ↑ flexibility	- ↑ tensile strength
- ↑ water sensitivity	- ↑ water resistance	- ↑ water sensitivity	- ↑ water resistance
	- ↑ solvent resistance	- ↑ easy of solvation	- ↑ solvent resistance
	- ↑ adhesion to hydrophobic surfaces		- ↑ adhesive strength

It is not intended here to review all the above mentioned properties of PVAL. More details are given in the literatures [67, 73, 84, 85]. In this section, the solubility of PVAL in water will be discussed because PVAL matrix is extracted by water after melt spinning of PLA/PVAL blend.

The solubility of PVAL in water depends on its degree of polymerization and especially significant is the effect of hydrolysis. Commercial PVAL grades can be classified as fully hydrolysed (97.5-99.5 % degree of hydrolysis) and partially hydrolysed (87-89 %). Generally, PVAL grades with high degree of hydrolysis have low solubility in water (Figure 2.2). However, the solubility property of PVAL in water is complex. It also strongly depends on the solution temperature and treatment time [69].

Figure 2.3 represents the relation between the solubility and dissolving temperature of PVAL, with the degrees of polymerization ranging from 500 to 2400 and with various degrees of hydrolysis of ca. 98, 88, and 80 %. While the solubility of fully hydrolysed PVAL (98 % degree of hydrolysis) increases with the decrease of the polymerization degree, the solubility of partially hydrolysed PVAL (88 % degree of hydrolysis) is independent of the polymerization degree. The solubility of PVAL with the low degree of hydrolysis (80 %) shows a different phenomenon. It has a high solubility at low temperatures and drops as temperature increases. This phenomenon, known as the "cloud point", is well known with several water soluble polymers. To avoid the cloud point phenomenon, therefore, the partially hydrolysed grades PVAL are selected in considering their solubility at a temperature as low as possible [69, 86].

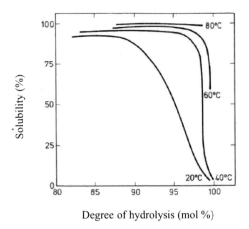

Figure 2.2 Solubility of a PVAL sample with the degree of polymerization of 1700 in the relation with the degree of hydrolysis at various dissolution temperatures (adapted from Ref. [69], with permission from John Wiley and Sons)

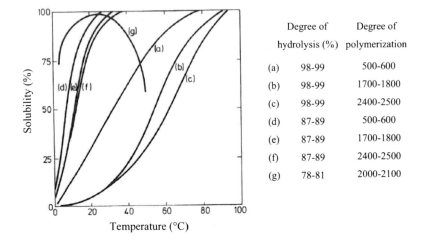

	Degree of hydrolysis (%)	Degree of polymerization
(a)	98-99	500-600
(b)	98-99	1700-1800
(c)	98-99	2400-2500
(d)	87-89	500-600
(e)	87-89	1700-1800
(f)	87-89	2400-2500
(g)	78-81	2000-2100

Figure 2.3 Solubility of various PVAL in water vs. solution temperature (adapted from Ref. [69], with permission from John Wiley and Sons)

Background and literature survey

2.1.3 Poly(lactic acid) (PLA)

Poly(lactic acid) (PLA), also known as polylactide, is a biodegradable thermoplastic aliphatic polyester. It is derived from 100 % renewable resources such as corn and sugar beets [87]. PLA can be performed either by both direct polycondensation of lactic acid, produced from sugar fermentation by bacteria or via ring-opening polymerization (ROP) of lactide (LA), a cyclic lactic acid dimer [88-91] (Figure 2.4). It is recorded that Pelouze first synthesized PLA by polycondensation of lactic acid in 1845 [92]. In 1932 Carothers developed a method to polymerize lactic acid into a low molecular weight PLA by heating lactic acid in vacuum [93]. This method was later patented by Du Pont in 1954 [94].

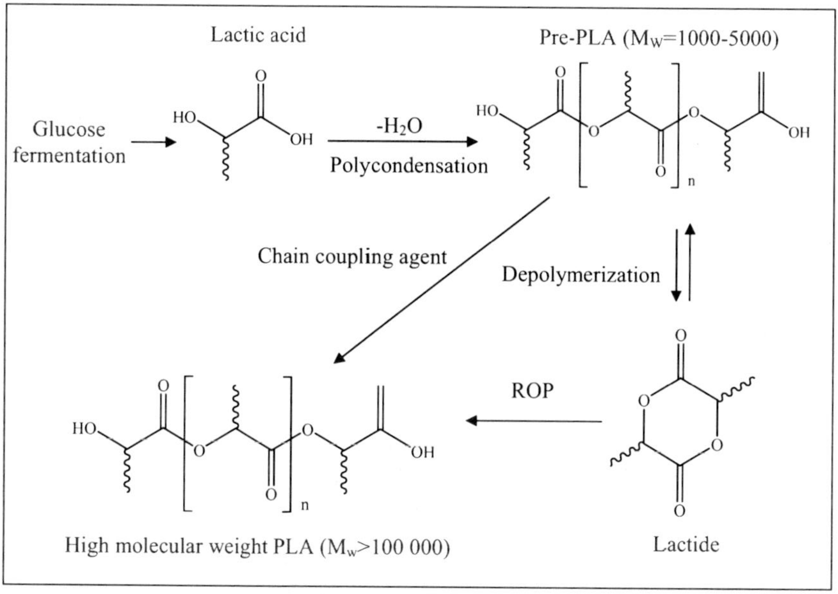

Figure 2.4 Reaction routes of producing PLA from lactic acid [63, 87, 95, 96]

PLA can be made into different resin grades, which can be used for various conventional melt processes such as injection molding, film and sheet casting, extrusion flow film, foaming, electrospinning, melt spinning, etc. [97]. The products of PLA have a large number of applications, which can be divided into major fields [96]: packaging, textile, plastic engineering, disposable ware, and medical applications. Nowadays, PLA is widely used as scaffolds for tissue engineering because of their biodegradable and biocompatible properties. It was reported that the micro/nanofibrous scaffolds successfully produced from PLA, PLA copolymer, or PLA blend with other materials. These scaffolds can be used for several kinds of tissues: *in vitro* culture of nerve stem cells [98, 99], bone tissue regeneration [100], heart [101], cartilage [102], and vascular graft [103-109].

2.2 Polymer blends with micro-nanofibrillar phase morphology

2.2.1 Polymer-polymer miscibility and phase separation

Polymer blends are generally classified into either miscible (completely mix on a molecular scale) or immiscible blends, which is the more often case, forming a multi-phase structure. Phase separation is also determined by thermodynamic relationships for the Gibbs free energy of mixing.

$$\Delta G_{mix} = \Delta H_{mix} - T \cdot \Delta S_{mix} \tag{2.1}$$

where ΔG_{mix} is the free energy of mixing, ΔH_{mix} is the enthalpy of mixing (heat of mixing) and ΔS_{mix} is the entropy of mixing. A necessary and sufficient condition for miscibility, i.e. a homogeneous phase from polymer-polymer blend, requires the free energy of mixing ΔG_{mix} to be negative and

$$\left(\frac{\partial^2 \Delta G_{mix}}{\partial \emptyset_i^2}\right)_{T,P} > 0 \tag{2.2}$$

where $\emptyset_i$ is the volume fraction of component i, T and P are the temperature and the pressure, respectively.

Most polymers are high molecular weight and their freedom of movement is limited along their molecular axes. In other words, the value of the combinatorial entropy of mixing ΔS_{mix} is negligibly small in such systems. On the other hand, it is generally agreed that the mixing of high molecular weight polymers is often an endothermic process, i.e. the heat of mixing is a positive $\Delta H_{mix} > 0$. The value of ΔG_{mix} is dominated by ΔH_{mix}. Consequently, most polymer blends are likely to be thermodynamically immiscible due to the positive Gibbs free energy of mixing $\Delta G_{mix} > 0$.

In general, the Gibbs free energy of mixing ΔG_{mix} is negative only if there are some strong specific interactions between the two polymers. The most common specific interactions found in polymer blends are hydrogen bonding, dipole-dipole, and ionic interactions. An overview of these interactions in polymer blends is available in the references [110-112].

2.2.2 Phase morphologies of immiscible polymer blends

As mentioned above, polymer blends are often immiscible systems consisting of multiple phases. Figure 2.5 gives different types of useful morphologies ranging from disperse drops to fibers to lamella to co-continuous structures for different end properties such as high strength and toughness, good barrier properties, etc. by a judicious melt blending process [113].

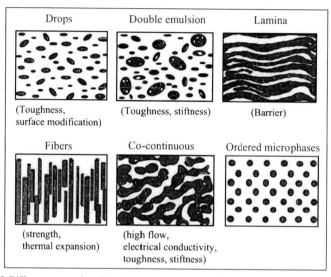

Figure 2.5 Different types of useful morphologies of immiscible polymer blends (reproduced from Ref. [113], with permission from John Wiley and Sons)

Generally, based on the degree of phase separation of two immiscible polymer blends, multiphase polymer blends can be broadly classified into two types: blends with discrete phase (droplets in matrix) and blends with a bicontinuous phase (co-continuous) [114, 115]. For a given blend component and under appropriate processing conditions, the dispersed phase can be deformed in micro- and nanofibrillar phase morphology in the matrix, which are often called matrix-fibrillar structures. From the view point of polymer composites, these structures are named *in situ* reinforced polymer-polymer composites or micro-/nanofibrillar composites (MFCs/NFCs) by Fakirov and Bhattacharyya et al. [34, 116, 117]. These MFCs produced from polymer blends can be improved their mechanical properties, especially tensile strength and modulus [118-121], thermal stability [122, 123] and their dyeability. The MFCs and/or NFCs open the opportunity to prepare micro- or nanofibrils if the matrix phase can be easily removed by appropriate solvents.

2.2.3 The formation of micro-nanofibrillar morphology in an elongational flow

It is well established that when blending two immiscible polymers, the formation of polymer blends, that results in micro- or nanofibrillar phase morphology, depends on the properties of blend components (e.g. the blend ratios, viscoelasticity of components, interfacial tension between components) as well as on the processing conditions (temperature, draw ratio, shear rate, shear stress, total strain, etc.). Under certain conditions, the micro- and nanofibrillar structures can be formed during melt processing by different methods, such as injection molding, sheet extrusion, film extrusion and fiber melt spinning.

Among the above mentioned methods, melt spinning is the most effective method to form a dispersed phase into fibrillar structures (as also known fiber-like morphologies) in the matrix of polymer blends due to the presence of a combination of shear and elongational flow fields. Furthermore, the elongational flow occurring in the melt spinning process seems to be more effective in producing the fibrillar morphology than an only shear flow field existing in other processing methods due to extreme changes in elongation rate [24, 124-126].

In the melt spinning process, polymer blends mainly undergo shear flow during compounding in single or twin-screw extruder and also during passing the capillary in the spinneret. After that, the elongational deformation along the spinline becomes more important [127-130]. Numerous studies on morphology development of polymer blends in single- and twin-screw extruder as well as in the spinneret channel, and the effect of various parameters on it have been published [14, 40-54]. By contrast, few studies have tried to study the morphological variations of polymeric blend systems after extrusion from spinneret orifices into fibers [26, 55, 56]. Padsalgikar and Ellision [55] developed a model for simulation of droplet deformation to predict the morphology of polypropylene (PP)/polystyrene (PS) blend fibers from polymer blends. They studied the effect of take-up speed on the droplet deformation. It was reported that with an increase in the take-up speed, there was a linear increase in the maximum strain rate, while the droplet diameter did not decrease linearly. Yang et al. [56] studied the phase morphology development of polypropylene (PP)/ethylene-butene copolymer (EMB) with a low interfacial tension between them. They concluded that elongational stress can stretch the droplets into long fibrils. Interfacial tension can not break up these fibrils because the molten polymer blends were solidified too fast, while coalescence takes effect to merge PP fibrils at fiber surface together to form a continuous PP surface layer. Recently, Tavanaie et al. [26] studied the effect of blend ratio on the morphological properties of blend fibers. The morphology of dispersed phase was also evaluated by rheological analyses. These investigations [26, 55, 56] focused on the morphology of the dispersed phase in blend samples at only two positions along the spinline: Blend samples at the die exit and the as-spun fibers taken from the bobbin. These studies showed that no fibrillar morphology of the dispersed phase was found in the blend extrudates (undrawn filaments) at the die exit, while continuous fibrillar structures were observed in the as-spun blend fibers (blend filaments). In our recent study [36], biodegradable PLA nanofibrillar structures were successfully fabricated by using the conventional melt spinning method. We also found that there was a significant difference of morphology of the PLA dispersed phase in PLA/PVAL blend extrudates and in PLA/PVAL blend filaments. While the PLA structures in PLA/PVAL blend extrudates showed the shape of spherical or ellipsoidal droplets with typical diameters in microscale (~1 − 5 μm), the PLA structures in PLA/PVAL blend filaments appeared as uniform continuous long thin fibrils, their typical diameters here were in nanoscale (~30 − 200 nm). It is well known that extreme changes in elongation rate associated with tension stress within fiber formation zone strongly affected the morphology development of the dispersed phase in polymer blends along the spinline.

In the melt spinning process, the diameter $D(x)$, the velocity $v(x)$, the temperature $T(x)$, the tensile force $F(x)$, tensile stress $\sigma(x)$ and the apparent elongational viscosity $\eta_{app}(x)$ vary along the spinline. These parameters play an important role in the development of fiber morphology and help to explain the fibrillation process of the dispersed phase in polymer blends. In principle, the final morphology is the result of a balance between deformation-disintegration phenomena (drop breakup) and coalescence [126]. Therefore, it is necessary to consider all factors (the changes of filament parameters as well as drop breakup and coalescence phenomena along the spinline) that may explain the mechanism of the morphological development of polymer blends from rod-like to nanofibrillar structures.

2.2.4 Theoretical considerations of droplet deformation and break-up

Droplet deformation has been intensively investigated both experimentally and theoretically since the pioneering work by Taylor [131, 132]. In principle, droplet of dispersed phase in the matrix undergoes shear and elongational flow. The droplet deformation in simple shear flow ($v_x = \dot{\gamma}y$, $v_y = 0$, $v_z = 0$; $\dot{\gamma}$ is the shear rate) has been studied using two parallel plates, in plane hyperbolic flow ($v_x = \dot{\gamma}x$, $v_x = \dot{\gamma}y$, $v_z = 0$) using four-rollers (Figure 2.6) [132, 133].

Background and literature survey

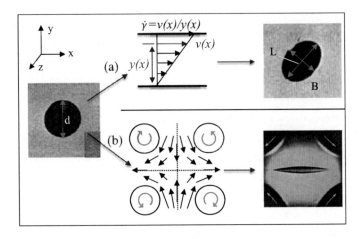

Figure 2.6 Droplet deformation in simple shear (a) and in plane hyperbolic flow fields (b), modified after [132]

During shear flow, the droplet deformation process of Newtonian system is mainly governed by the dimensionless capillary number Ca and viscosity ratio λ, defined as follows:

$$Ca = \frac{\text{hydrodynamic stress}}{\text{interfacial stress}} = \frac{\eta_m \dot{\gamma} R}{\Gamma} \qquad (2.3)$$

$$\lambda = \frac{\eta_d}{\eta_m} \qquad (2.4)$$

where η_d and η_m are the viscosity of the dispersed phase and the matrix, respectively, $\dot{\gamma}$ is the shear rate, $R = d/2$ is the undeformed droplet radius, and Γ is the interfacial tension.

For other flow fields such as uniaxial elongational flow, plane hyperbolic flow, etc., Ca can be defined in analogy to Equation 2.3 using the relevant component of the rate deformation tensor. The droplet deformation D is given by [134]

$$D = \frac{L - B}{L + B} \qquad (2.5)$$

where L and B are the length and breadth of deformed droplets, respectively.

The droplet deformation was calculated by Taylor [131, 132] using the perturbation method with spherical shape as a zero approximation. The droplet deformation was quantitated in the two Equations 2.6 and 2.7:

$$D = Ca \cdot \frac{19\lambda + 16}{16\lambda + 16} \qquad (2.6)$$

if the interfacial tension effect dominates over the viscous effect ($Ca \ll 1$) and

$$D = \frac{5}{4\lambda} \qquad (2.7)$$

when the interfacial tension is negligible compared to the viscous effect ($\lambda \gg 1$).

It is worth noting here that Taylor's theory was limited in two above cases where either the interfacial tension effect is dominant over the viscous effect or vice versa in uniform shear and plane hyperbolic flow. The droplet deformation of the Taylor's theory was extended by Cox [135] to systems with the full range of viscosity ratios for steady shear flow (Equation 2.8),

$$D = \frac{5(19\lambda + 16)}{4(\lambda + 1)\left[(19\lambda)^2 + \left(\frac{20}{Ca}\right)^2\right]^{1/2}} \qquad (2.8)$$

for steady plane hyperbolic flow (Equation 2.9),

$$D = 2.0 \cdot Ca \cdot \frac{19\lambda + 16}{16\lambda + 16} \qquad (2.9)$$

and for steady axisymmetric extensional flow ($v_x = -(\dot{\gamma}/2)x$, $v_x = -(\dot{\gamma}/2)y$, $v_z = \dot{\gamma}z$) (Equation 2.10)

$$D = 1.5 \cdot Ca \cdot \frac{19\lambda + 16}{16\lambda + 16} \qquad (2.10)$$

The droplet will breakup into smaller droplets when the capillary number reaches its critical value. The critical capillary number Ca_c can be determined from experimental studies. Droplet deformation and breakup depend on the reduced capillary number Ca^*, defined as

$$Ca^* = \frac{Ca}{Ca_c} \qquad (2.11)$$

Depending on the Ca^* value, the droplets will either deform or break based on the following categories:

- If $Ca^* < 0.1$, there is no deformation of droplets.

Background and literature survey

- If $0.1 < Ca^* < 1$, there is no breakup.
- If $1 < Ca^* < 4$, there is a droplet deformation, but they break conditionally.
- If $Ca^* > 4$, droplets affinely deform with the rest of the matrix and extend into long stable filaments. The diameter, B, and length, L of droplet can be calculated using the affine deformation theory:

$$B = B_0 exp\left(-\frac{\varepsilon}{2}\right) \quad (2.12)$$

$$L = L_0 exp(\varepsilon) \quad (2.13)$$

where B_0 and L_0 are the diameter and the length of the original droplets, respectively, the strain $\varepsilon = \dot{\varepsilon}t$ is the product of the elongational rate $\dot{\varepsilon}$ and the time t.

The droplet deformation for $0.1 < Ca^* < 4$, is calculated by the above deformation Equations 2.8, 2.9, and 2.10.

In the case Ca^* between 1 and 4, the time t_b^* required for the break up can be calculated using the following empirical equation [136]:

$$t_b^* = 84 \, \lambda^{0.345}(Ca^*)^{-0.559} \quad (2.14)$$

Break-up does not occur if $t < t_b^*$.

Many experimental studies have shown that Ca_c is a function of viscosity ratio λ. De Bruijn [137] found that droplets break in shear flow most easily when $0.1 < \lambda < 1$, but the drops do not break up when $\lambda > 4$. In the case < 4, the Ca_c for droplet deformation can be expressed as a function of λ

$$\log(Ca_c/2) = c_1 + c_2 \log\lambda + c_3(\log\lambda)^2 + c_4/(\log\lambda + c_5) \quad (2.15)$$

where, $c_1 = -0.5060$, $c_2 = -0.0994$, $c_3 = -0.1240$, $c_4 = -0.1150$, $c_5 = -0.6110$.

The four different breakup mechanisms in simple shear flow were summarized by Sundararaj depending on the viscosity ratio λ [138]:

- If $0.5 < \lambda < 9$, the drop may form a sheet in the flow direction and break up.
- If $0.05 < \lambda < 60$, the drop may slowly erode at the surface.
- If $\lambda \sim 7.5$, the drop may stretch in the vorticity direction and break up.
- If $0.05 < \lambda < 3$, the drop may spit out small droplets via a tip streaming mechanism.

In planar extensional flow, Grace [139] showed that the experimental results of drop deformation can well fitted using the above Equation 2.15 with $c_1 = -0.6485$, $c_2 = -0.0244$, $c_3 = -0.0221$, $c_4 = -0.00056$, $c_5 = -0.00645$. The author plotted Ca_c for droplet break up in uniform steady shearing and in planar hyperbolic flow over viscosity ratios as shown in Figure 2.7. It is indicated that elongation is more efficient than shear for breaking the drops.

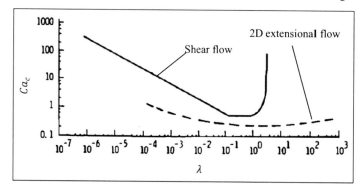

Figure 2.7 Critical capillary number Ca_c vs. viscosity λ ratio for droplet breakup in shear and hyperbolic flow fields, adapted from Ref. [139]

2.3 Theoretical considerations of thin filament model in the melt spinning

The melt spinning process for the manufacturing of polymeric fibers is well known since 1930's. By using this process, the first successful melt spun nylon 66 (PA66) fiber was developed by Carothers et al. [140, 141]. In the melt spinning process, polymer pellets or granules are fed into an extruder, usually in single-screw extruder, melted by heat, and then the molten polymer is pumped through a spinneret (die) under pressure. The molten extrudates are then quenched by cooling air, drawn down to a smaller diameter, solidified into filaments and taken-up by a winding device. Figure 2.8 shows a schematic diagram of the melt spinning process.

Since the late 1950's, many theoretical and practical investigations of the dynamics of the melt spinning process were performed by Ziabicki et al. [142-152], Kase and Matsuo [153-158], Han et al. [127, 159-166], Andrews [167], Hamana [168], George [169], Lin [170]. In these studies the basic differential equations were developed that allows one to simulate the dynamics of the fiber formation in the melt spinning process. The mathematical models of the dynamics of the fiber formation in the melt spinning process are often described by using the thin filament model, because it is based on the purely extensional flow field with an

Background and literature survey

assumption: uniform distribution of axial velocity and temperature across the filament. In other words, radial distribution of axial velocity and temperature are neglected [171]. The thin filament model is based on the further simplifying assumptions: vertical spinline, the steady and incompressible melt flow (extrudate/filament).

The following sections review the basic balance equations of the thin filament model of the melt spinning process that describe the deformation, cooling, and stress development processes of filaments. More complete and accurate details of mathematical modeling were discussed and reviewed in four monographs by Ziabicki in 1976 [172], Ziabicki and Kawai in 1985 [173], Nakajima et al. in 1994 [174], and recently by Beyreuther and Brünig in 2007 [175].

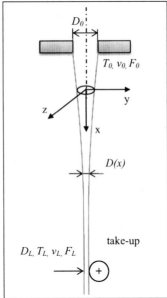

Figure 2.8 A schematic diagram of fiber formation in melt spinning process

2.3.1 Mass balance equation

The mass balance presents the continuity equation of the melt spinning process. The continuity equation is expressed as

$$Q = \varrho_P(x) \cdot A(x) \cdot v(x) = \text{constant} \; or \tag{2.16}$$

$$Q = \varrho_P(x) \cdot \frac{\pi}{4} D(x)^2 \cdot v(x) \tag{2.17}$$

where, Q is the mass flow rate, $\varrho_P(x)$, $v(x)$, $D(x)$, and $A(x)$ denote, respectively, the mass density of polymer, the velocity of filament, the circular cross-sectional diameter, and the cross-sectional area of filament at a distance x from the spinneret.

In textile engineering, the product of the filament cross-sectional area and its density is known as the fineness of filament T_t

$$\varrho_P \cdot A = \varrho_P \cdot \frac{\pi}{4} D^2 = T_t \tag{2.18}$$

The fineness T_t quantifies the linear mass density, i.e. mass per unit length of the filament (=titre) and has its own special units. The most common units are *decitex* (abbr. *dtex*) and *denier* (abbr. *den*). 1 dtex and 1 den are defined as the mass of 1 gram per 10000 meters and per 9000 meters, respectively. It is seen from the Equation 2.18 that the filament diameter D is proportional to the square root of its linear mass density. The relation between the fineness T_t (in dtex or den) and diameter of filament D (in µm) depends on the mass density ϱ_P (in g/cm^3) and is given by [175]

$$T_t = 0.0078\, \varrho_P D^2, \qquad D = 11.3\sqrt{T_t/\varrho_P} \text{ for } T_t \text{ in dtex} \tag{2.19}$$

$$T_t = 0.0112\, \varrho_P D^2, \qquad D = 9.44\sqrt{T_t/\varrho_P} \text{ for } T_t \text{ in den} \tag{2.20}$$

2.3.2 Force balance equation

Force balance equation can be used to determine the forces acting on the filament and was introduced for the first time by Ziabicki [146]. The force contributions are schematically shown in Figure 2.9 and expressed as following equations [175]:

$$F(x) = F_{rheo}(x) = F_L - F_{grav}(x) + F_{surf}(x) + F_{aero}(x) + F_{inert}(x) \text{ or} \tag{2.21}$$

$$F(x) = F_{rheo}(x) = F_0 + F_{grav}(x) - F_{surf}(x) - F_{aero}(x) - F_{inert}(x) \tag{2.22}$$

where, F_L is the take-up force, applied to the filament by the take-up device, F_0 is the initial force at the capillary exit, $F_{surf}(x)$ is the surface tension force, $F_{grav}(x)$ is the gravitational force, $F_{aero}(x)$ is the air drag force and $F_{inert}(x)$ is the inertial force.

Background and literature survey

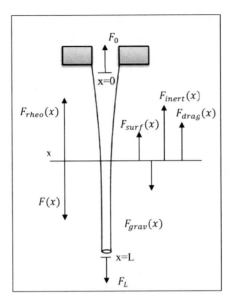

Figure 2.9 A schematic draw of forces acting on a filament, adapted from Ref. [175]

These forces at any distance x from the spinneret can be expressed as follows:

$$F_{grav}(x) = \int_x^L \varrho_p \cdot g \cdot \dot{V}/v(\tilde{x}) \, d\tilde{x} \tag{2.23}$$

$$F_{surf}(x) = \pi \cdot \sigma_{surf} \cdot (R_0 - R(x)) \tag{2.24}$$

$$F_{aero}(x) = 2\pi \int_0^x R(\tilde{x}) \cdot \frac{\varrho_{air}}{2} \cdot v^2(\tilde{x}) \cdot c_f(\tilde{x}) \cdot d\tilde{x} \tag{2.25}$$

$$F_{inert}(x) = Q \cdot (v(x) - v_0) \tag{2.26}$$

where, $\dot{V}$ is the volumetric flow rate, v is the velocity of filament, σ_{surf} is the surface tension of the material, R_0 is the initial radius of filament, ϱ_{air} is the mass density of air, and c_f is the friction coefficient of air. The air friction coefficient c_f describes the friction-caused momentum transfer between filament surface and ambient air. The air friction coefficient c_f depends on the state of air flow surrounding the filament within fiber formation zone. A numerous theoretical analysis and experimental investigations on air friction forces acting on the running filament have led to the following equation:

$$c_f = a' \cdot Re^{c'} \tag{2.27}$$

with different constant parameters a' and c', which depend also on the arrangement of measurements. The most common values $a' = 0.37$ and $c' = -0.61$ are used that suggested by Hamana [168]. A summary of different formulas with different parameters a' and b' can be found in the two monographs [173, 175]. Re denotes the non-dimensional Reynolds number. The Reynolds number itself is given by

$$Re(x) = \frac{\varrho_{air} \cdot v(x) \cdot 2R(x)}{\eta_{air}} \tag{2.28}$$

where, η_{air} is the dynamic viscosity of air.

2.3.3 Energy balance equation

The energy equation of a running filament in a cooling medium, usually ambient air, with temperature T_{air} is given by

$$\frac{dT(x)}{dx} = -(T(x) - T_{air}(x)) \cdot N_u \frac{\pi \cdot \lambda_{air}}{Q \cdot c_P} + \frac{\Delta H}{c_p} \cdot \frac{dX_c}{dx} \tag{2.29}$$

where, $T_{air}(x)$ is the temperature of surrounding air, λ_{air} is the heat conductivity of air, $Q = \varrho_P \cdot \dot{V}$ is the mass flow rate of polymer in which ϱ_P is the mass density of the polymer blend, c_P represents the specific heat capacity of the polymer, ΔH is the heat of fusion and X_c is the crystallinity degree. N_u is the Nusselt number, characterizing the heat transfer from filament surface into the ambient air.

Heat transfer from the melt filament to the ambient air involves the following mechanisms: radiation, free convection and forced convection [172, 175]. *The heat radiation* strongly depends on the temperature (power law with T^4 dependence). The heat radiation is only important wherever very high temperatures (more than ca. 500 °C) are involved. In polymer melt spinning with spinning temperatures of up to 300 °C, the radiation contribution is often negligible compared with the convective heat transfer. *The free (natural) convective heat transfer* involves stationary systems. The Nusselt number Nu for free convection can be written as follows [175]:

$$Nu_n = Nu_n(Gr, Pr) \qquad (2.30)$$

where, Gr is the Grashof number is given by

$$Gr = g\beta_{air}(T - T_{air})D^3 \qquad (2.31)$$

where, β_{air} as the air thermal expansion coefficient, D is the diameter of the filament. The Prandtl number of air is given by $Pr = \eta_{air}C_{p,air}/\lambda_{air}$, with η_{air} as the dynamic viscosity of air. *The forced convective transfer* becomes a dominant factor for the high speed spinning process. The Nusselt Number for the forced convection heat transfer depends on three dimensionless numbers: Reynolds (Re), Prandtl (Pr), and dimensionless distance (x/D), which can be described as follows [175, 176]:

$$Nu = Nu(Re_\parallel, Re_\perp, x/D, Pr) \qquad (2.32)$$

where, $Re_\parallel$ and $Re_\perp$ are the Reynolds number related to the parallel and cross air flow, defined as follows:

$$Re_\parallel(x) = \frac{v_\parallel(x) \cdot D(x)}{v_{air}} \qquad (2.33)$$

$$Re_\perp(x) = \frac{v_\perp(x) \cdot D(x)}{v_{air}} \qquad (2.34)$$

where, $v_\parallel(x)$, $v_\perp(x)$ is the axial, cross air velocity between filament and ambient air, respectively, v_{air} is the kinematic viscosity of air. It is given by $v_{air} = \eta_{air}/\varrho_{air}$, where η_{air} is the dynamic viscosity of air and ϱ_{air} is the mass density of air.

A numerous theoretical and experimental investigations have been developed to describe the relationship between the Nusselt and Reynolds numbers, which can be expressed in general equation:

$$Nu = a\left(Re_\parallel^2 + b \cdot Re_\perp^2\right)^c \qquad (2.35)$$

Table 2.2 gives the selected relationship between the Nusselt and Reynolds number for the melt spinning process from different literatures [153, 167, 176-181].

Table 2.2 Nusselt vs. Reynolds number relations, modified after [175]

Nu vs. Re number	References	Materials and spinning conditions
1) For only parallel flow: $Nu = a \cdot Re_{\parallel}^{2c}$ with		
1.1) a=0.764; c=0.19	Andrews [167]	- Material: PET - T_0(*): 280 °C - v_L(*): 33.6-4000 ft·min^{-1} (~10-1200 m·min^{-1}) - $\dot{V}$(*): 1.95 cm^3·s^{-1} (~117 cm^{-1}·min^{-1})
1.2) a=0.42; c=0.167	Kase and Matsuo [153]	- Materials: PET and poly(ethylene terephthalate-co-isophthalate) - T_0: 280 and 290 °C - v_L: 600 and 694 m·min^{-1}
1.3) a=0.325; c=0.15	Glicksman [177, 178]	- Material: Glass fiber - T_0: 2240 °F (~1230 °C) - v_L: 21.3-84.9 ft·s^{-1} (~390-1550 m·min^{-1})
1.4) a=0.16; c=0.26	Ohkoshi et al. [179, 180]	- Material: PEEK - T_0: 400 °C - v_L: 50-200 m·min^{-1} - Q(*): 3.5 and 7.1 g·min^{-1}
1.5) a=1.5; c= - 0.11	Golzar [176]	- Material: PEEK - T_0: 385 °C - v_L: 25, 1000, and 2000 m·min^{-1} - Q: 0.456, 2.0, and 3.0 g·min^{-1}
2) For parallel and transverse flow: $Nu = a\left(Re_{\parallel}^2 + b \cdot Re_{\perp}^2\right)^c$ with		
a=0.42; b=64; c=0.167 a=0.28; b=1024; c=0.17 a=0.33; b=4096; c=0.2	Kase and Matsuo [153] Brünig et al. [181]	- Materials: PET and PA6 - T_0: 280 and 260 °C - v_L: 1000- 4000 m·min^{-1} - Q: 1.0 and 1.4 g·min^{-1}

(*) T_0: Extrusion temperature, v_L: take-up velocity, $\dot{V}$: Volumetric flow rate, Q: Mass flow rate

Some above mentioned equations are combined to online measurement results of filament temperature (section 3.3.6) and filament velocity (section 3.3.5) are used to characterize all filament parameter profiles that will be presented in the section 4.3.

Chapter 3
Experimental

3.1 Materials

The PLA and the PVAL were used as dispersed phase and matrix material, respectively. Both materials are considered as semi crystalline polymers. They are suitable for the melt spinning process.

3.1.1 Poly(vinyl alcohol) (PVAL)

The PVAL (type Mowiflex TC 232, kindly donated by Kuraray® Europe GmbH, Germany) is the commercial thermoplastic polymer. The PVAL can be used in all common thermoplastic processes, including blow film extrusion and injection molding [182]. The PVAL Mowiflex TC 232 with the hydrolysis degree of 88 % and the polymerization degree of 1400 is water soluble at room temperature and biodegradable polymer with the specialties as summarized in Table 3.1.

Heat flow curves of PVAL

Figure 3.1 shows the heat flow curves (or DSC curves) of PVAL obtained using Differential Scanning Calorimetry (DSC)[1] in the temperature range from 0 °C up to 220 °C with a heating rate of 10 K·min^{-1}. For the second heating run (red curve), the glass transition temperature $T_{g,PVAL}$ and the melting transition temperature $T_{m,PVAL}$ are found to be around 34 °C and 173.9 °C, respectively. A cold crystallization process is not observed in the second heating run. PVAL was crystallized during cooling with $T_{c,PVAL}$ of 130.8 °C (blue curve). It means that crystalline part in PVAL has completely crystallized during cooling run. In principle, it can be concluded that PVAL is semi crystalline material because DSC curves displayed both

---

[1] The DSC experiment is presented in the section 3.3.2

Experimental

the glass transition and melting transition temperature, which are presented for the amorphous fraction and the crystalline domains, respectively.

Table 3.1 PVAL Specialties

Properties of PVAL	Results and testing conditions		
Molecular weight (g·mol^{-1})	67000[a]		
Mass density (g·cm^{-3})	1.27, after Ref. [183]		
Melting temperature (°C)	178[a]	174[b]	2nd heating
Glass Transition Temperature (°C)	35[a]	34[b]	2nd heating
Melt flow rate (g·10 min^{-1})	24.0-40.0[a]	38.8[c]	190 °C, 21.6 kg
		53.2[c]	195 °C, 21.6 kg
		69.9[c]	200 °C, 21.6 kg
		117.3[c]	210 °C, 21.6 kg
Zero-shear viscosity (Pa·s)		53423.4[d]	190 °C
		5037.9[d]	195 °C
Solubility in water			Room temperature
Degree of hydrolysis (mol-%)	88[a]		

[a] provided by Kuraray® [184],
[b] measured using Differential Scanning Calorimetry (DSC) presented in the section 3.3.2
[c] measured using the Melt Flow Index (MFI) technique presented in the section 3.3.4
[d] measured using oscillatory rheometer presented in the section 3.3.4 and fitted using Carreau model [185].

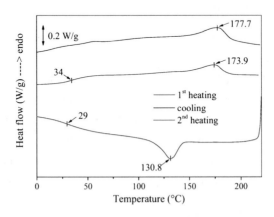

Figure 3.1 Heat flow curves of PVAL

3.1.2 Poly(lactic acid) (PLA)

The thermoplastic PLA (type Ingeo 6202D, 98 % L-Lactic) is the commercial melt spinning grade purchased from NatureWorks™ LLC, USA. Its melt-flow index (MFI) (210 °C, 2.16 kg), reported by NatureWorks LLC is 15 to 30 gram per 10 min, glass transition temperature and melt temperature ranging 55-60 °C and 160-170 °C, respectively. Table 3.2 summarizes some relevant specialties of PLA.

Heat flow curves of PLA

Figure 3.2 plots the heat flow curves of PLA granules in the temperature range from 0 °C up to 220 °C with a heating rate of 10 K·min^{-1}. Similarity with PVAL, DSC curves of PLA exhibited both glass transition temperature $T_{g,PLA}$ and melting transition temperature $T_{m,PLA}$. PLA is also known as high crystalline thermoplastic polymer. The maximum crystallinity of used PLA 6202D is known ranges from 40 to 50 % [186]. It has been widely used for the melt spinning process.

Table 3.2 PLA Specialties

Properties of PLA	Results and testing conditions		
Molecular weight (g·mol^{-1})	157000[a]		
Mass density (g·cm^{-3})	1.24[a]		
Melting temperature (°C)	160-170[a]	160/167[b]	2nd heating
Glass Transition Temperature (°C)	55-60[a]	61[b]	2nd heating
Melt flow rate (g·10 min^{-1})		9.4[c]	190 °C, 2.16 kg
		11.5[c]	195 °C, 2.16 kg
		14.1[c]	200 °C, 2.16 kg
	15.0-30.0[a]	22.1[c]	210 °C, 2.16 kg
Zero-shear viscosity (Pa·s)		1121.09[d]	190 °C
		358.76[d]	195 °C

[a] reported by NatureWorks™ [186],
[b] measured using Differential Scanning Calorimetry (DSC) presented in the section 3.3.2
[c] measured using the Melt Flow Index (MFI) technique presented in the section 3.3.4
[d] measured oscillatory rheometer presented in the section 3.3.4 and fitted using Carreau model [185].

Experimental

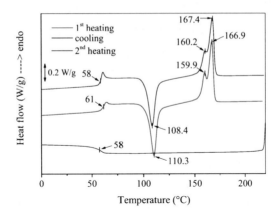

Figure 3.2 Heat flow curves of PLA

3.2 Processing

3.2.1 Melt mixing

For mixing the PLA/PVAL systems, three different types of mixing devices are used: microcompounder, internal mixer, and twin screw extruder. Before starting the mixing processes, the PVAL and PLA were dried at a temperature of 60 °C for 8 h and 80 °C for 6 h, respectively.

Microcompounder

The PLA was melt-compounded with the PVAL in different weight ratio ranging from 10/90 to 50/50 using a co-rotating twin-screw microextruder (15 mL microcompounder, DSM Xplore, Geleen, The Netherlands). The mixing temperature was set at 195 °C with 100 rpm screw speed for 5 min. The PLA/PVAL blends produced by this device were used for investigations of their thermal, rheological properties (section 4.2.1), and their morphological properties (section 4.2.3).

Internal mixer

The PLA/PVAL blend pellets at weight ratio of 30/70 were melt compounded using a C. W. Brabender laboratory mixer (Mixer W50, Brabender® GmbH & Co. KG, Duisburg, Germany). The mixing temperature was set at 195 °C and the rotor speed was held constant at

70 rpm. After the mixing period of 5 min, the PLA/PVAL molten blend was taken from the chamber and compressed into thin plates, which were then pelletized and used for the melt spinning process on a piston spinning device.

Twin screw extruder

The PLA/PVAL blend pellets at weight ratio of 30/70 were also mixed and extruded using a HAAKE Poly Lab OS twin screw extruder (PTW 16/25, Thermo Fisher Scientific Germany BV & Co KG, Braunschweig, Germany). The mixing temperatures at different zones along the extruder (starting from the feeding zone to the die) were 175 °C, 180 °C, 185 °C, 190 °C, 190 °C, and 180 °C. The screw speed was 100 rpm with the feeding rate of 1000 grams per hour. The mixed extrudates were then cut into small pellets. These pellets were melt spun also as PLA/PVAL monofilament on a piston spinning device.

3.2.2 Melt spinning

The melt spinning experiments were carried out using two different types of spinning equipment: piston- and extruder melt spinning device.

Piston melt spinning device (small-scale melt spinning device)

Piston type spinning device, which was constructed by our own machine shop at the IPF Dresden, was used for spinning of small polymer quantities from different kinds of spinnable polymers. All experiments can be performed over mass flow rates ranging from 0.01 $g \cdot min^{-1}$ up to 5 $g \cdot min^{-1}$ and take-up speeds from 5 to 2000 $m \cdot min^{-1}$. The melt spinning process on the piston type spinning device is often also known as pre-spinning step that are necessary to test the spinnability of polymer and control the spinning parameters before polymers will be prepared on the extruder spinning device. In this study, the piston spinning device was used to carry out a numerous spinning experiments for capturing the PLA/PVAL blend filaments (section 3.3.8), which are further used for the investigation of the morphological properties (section 4.3 and 4.4). This device was also used to perform a lot of spinning experiments for online measurements of filament velocity (section 3.3.5) and filament temperature (section 3.3.6), which are further employed for the characterization of PLA/PVAL 30/70 monofilaments profiles (section 4.3).

Before melt spinning, the PLA/PVAL pelletized blends (obtained by internal mixer or twin-screw extruder, section 3.2.1) were dried in a vacuum oven at a temperature of 60 °C for 6 h

Experimental

and then filled into the cylinder under dry nitrogen. The PLA/PVAL pelletized blend was melted in a heatable cylinder and pressed through a single hole die by a piston. All spinning experiments were performed at 195 °C. The diameter and the length of the capillary hole was $D_0 = 0.6$ mm and $L = 1.2$ mm, respectively, (aspect ratio $L/D_0=2$), the entrance angle of the capillary was $\alpha = 60°$. The take-up speed is altered from v=10 m·min^{-1} to 70 m·min^{-1} at a constant volumetric flow rate of $\dot{V}=0.78$ cm^3·min^{-1} (Q=1.0 g·min^{-1}) and the flow rate is also varied at a constant take-up of velocity v=50 m·min^{-1}. The melt-spun monofilament was collected on a winder, located 200 cm below the die exit.

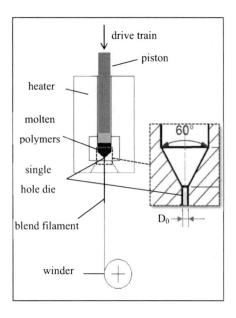

Figure 3.3 Schematic drawing of piston type extrusion spinning equipment

Extruder melt spinning device (industrial-scale melt spinning device)

Melt spinning of PLA/PVAL 30/70 multifilament was carried out on industrial-scale spinning equipment used for the manufacturing of synthetic fibers at the IPF Dresden (Figure 3.4). The spinneret contains 24 holes and the diameter of each capillary hole was 0.6 mm with an aspect ratio (L/D) of 4.

Before melt spinning, the dried PLA pellets were dry-mixed with PVAL in the weight ratio PLA/PVAL 30/70 in a vacuum oven at a temperature of 40 °C for 16 h and then transferred to a hopper that was purged with dry nitrogen. The extruder profile was 175, 180, 185, 190, and 195 °C (starting from the feeding zone to the die). Three spinning speeds were used 30, 40, and 50 m·min^{-1} with the constant throughput rate of 20.5 g·min^{-1}. In this way, filament yarns of 683, 512, 410 tex 24f, respectively were obtained.

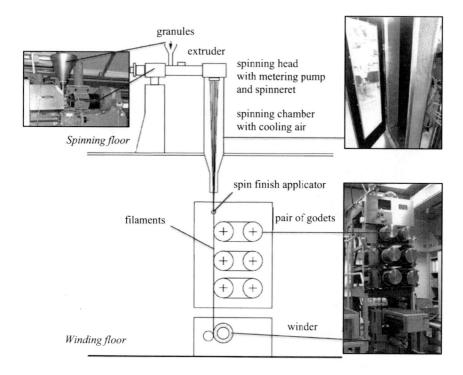

Figure 3.4 Principle of melt spinning device in IPF Dresden [175]

3.2.3 Off-line drawing process

The drawing process is achieved by stretching the filaments between two rollers (input and first output feed roller) with the 1st output feed roller rotating faster (Figure 3.5). The filaments are supplied at the input feed roller with a constant feed velocity, V_{in}, and are stretched by the 1st output feed roller running with a constant drawing velocity, V_{1out}, equaling $DR \times V_{in}$; where DR is the nominal draw ratio. For this study, the draw speed (the velocity of the winder), $V_{take-up} = V_{1out} = V_{2out}$ was kept at 30 m·min^{-1}, while the feed speed (the velocity of the input feed roller), V_{in}, was continuously adjusted in order to achieve the desired draw ratio.

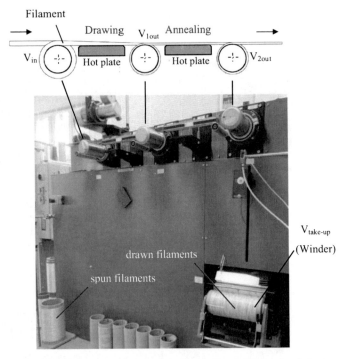

Figure 3.5 Scheme of the drawing process (above) and drawing device at IPF Dresden (below)

3.3 Characterization and on-line measurements

3.3.1 Attenuated total reflection-Fourier transform infrared (ATR-FTIR) spectroscopy

A Fourier transform infrared spectrometer Vertex 80v (Bruker Corporation, USA) were used to characterize the presence of specific chemical groups of polymer and interpolymer interaction. The spectra of the ATR-FTIR were recorded over the range of $4000 - 600$ cm^{-1}, 100 scans for each measurement with 4 cm^{-1} resolution using Mercury Cadmium Telluride (MCT) detector. The PLA, PVAL pellets, and their blends of ca. 0.5 g were pressed using a compression molding under the temperature of 185 °C, 185 °C, and 170 °C, respectively, to produce thin film with a thickness of ca. 50 μm.

3.3.2 Differential scanning calorimetry (DSC)

DSC was used to characterize the melting and crystallization behavior of both pure components and their blends. In this study, thermal analysis was carried out by DSC using a DSC Q1000 calorimeter (TA Instruments, Delaware, USA) under a nitrogen atmosphere. Calibration was performed with an indium standard. Specimens of about 5 mg were encapsulated in aluminum pans. Heating-cooling-heating scans were performed in a temperature range of -80 to 220 °C with a scan rate of 10 K·min^{-1}.

3.3.3 Thermogravimetric analysis (TGA)

All TGA analyses were conducted using a TGA Q5000 (TA Instruments) under a nitrogen atmosphere in a temperature range of 40-800 °C. Isothermal TGA measurements were also performed. The sample weight was around 5 mg in all experiments. After preheating at 40 °C for 5 min, the samples were heated up to 185, 190, 195, 200, and 205 °C with a heating rate of 100 K·min^{-1} and then examined isothermally for one hour.

3.3.4 Rheological measurements

Melt flow index (MFI)

Melt flow index (MFI) or Melt flow rate (MFR) of PVAL and PLA were measured using the Melt-Index Tester type MeltFlow @on plus (Karg Industrietechnik, Krailling, Germany) with nozzle length and diameter of 8 mm and 2.095 mm, respectively, at the testing conditions as shown in Table 3.1.

Table 3.1 Testing conditions of MFI

Conditions	PVAL	PLA
Temperature (°C)	190, 195, 200, 210	190, 195, 200, 210
Load (kg)	21.6	2.16
Time of preheating (s)	240	240

Shear viscosity using oscillatory rheometer

Rheological properties of the polymers were measured using an oscillatory rheometer ARES-G2 from TA Instruments. The specimens were measured using parallel plate disc geometry of 25 mm diameter with 2 mm gap between the plates. The frequency sweep experiments were performed at a strain amplitude of 1.25 % in a frequency range of 100 to 0.1 Hz at

Experimental

185 – 205 °C at an increment of 5 °C. Nitrogen gas was used to prevent thermal oxidation of the materials.

3.3.5 On-line filament speed measurement

The online measurements of the filament velocity were performed by using a laser Doppler anemometry device Laser Speed LS50M. It consists of optical sensor (model 250100), processor LS50MP, and a PC with configuration software TSI LaserTrak™ (version 3.4) manufactured by TSI Inc. based on technique known as Laser Doppler Velocimetry (LDV) [187] (Figure 3.6). This technique is a non-contact method for measuring the speed of moving solid state surface or particle velocity in a fluid by laser interferometry.

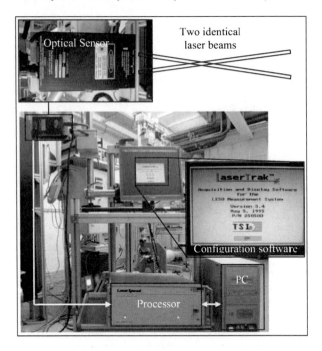

Figure 3.6 Model LS50M multiplexed LaserSpeed® system

The main principle of LDV is shown in Figure 3.7 and can be found in more details in references [188, 189]. The laser light is emitted from a laser diode through collimating lens. After the light is collimated, the beam is split into two identical beams that cross at an angle to create an interference pattern made of bright and dark fringes. The spacing between two adjacent bright fringes δ depends on the wavelength of laser light λ and the angle between the two beams θ. The area where the laser beams intersect, the velocity is measured.

The optical sensor collects the light scattered from particles moving through the intersection of the two coherent laser beams. The optical sensor focuses then this scattered light on a photo detector that converts the optical energy into electrical energy which is sent to a signal processor and converted to velocity data.

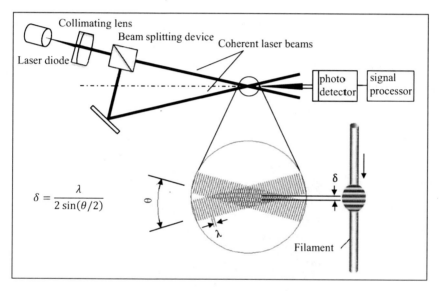

Figure 3.7 Layout of optical path to the filament, modified after [187]

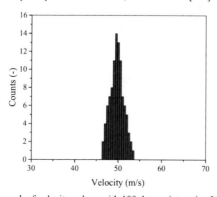

Figure 3.8 A histograph of velocity values with 100 data points using LDV technique

Experimental

In the present study, velocity recordings were obtained at distances in the range of 2.5 – 100 cm from the die exit, at each 2.5 cm or 5 cm intervals. A set of 5x100 data points was gathered for each distance from the die exit. A typical histogram of a velocity measurement with 100 data points obtained by using LDV technique is shown in Figure 3.8.

3.3.6 On-line filament temperature measurement

The filament temperature was measured using an infrared camera VarioTHERM™ with microscope lens MWIR/f4.4, designed by Jenoptik AG, Jena, Germany, combined with the monitoring subsystem including a personal computer (PC) and software package IRBIS® supported by InfraTec GmbH, Dresden, Germany (Figure 3.9). More details on this infrared thermography device are given by Golzar et al. [190].

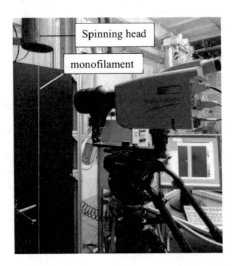

Figure 3.9 Photographic of test stand for temperature measurement using infrared camera

The infrared thermography method for measuring the filament temperature is based on using infrared radiation. Therefore, this method allows non-contact determination of filament temperature. The radiated energy received by the infrared camera E includes three components: emitted E_{em}, reflected E_{re}, and transmitted energy E_{tr} and can be expressed as follows:

$$E = E_{em} + E_{re} + E_{tr} \text{ or} \tag{3.1}$$

$$1 = \frac{E_{em}}{E} + \frac{E_{re}}{E} + \frac{E_{tr}}{E} \qquad (3.2)$$

The Equation 3.2 can be simplified as follows:

$$1 = em + re + tr \qquad (3.3)$$

where, em is the emissivity, which indicates the ability of infrared energy of filament is emitted to infrared camera, the reflectivity re and the transmissivity tr indicate the ability to reflect and transmit infrared energy, respectively.

For a perfect black body, the emissivity em is equal one; the reflectivity re and the transmissivity tr are zero ($em = 1, re = 0, and\ tr = 0$). Polymeric filaments are not perfect black bodies and the emissivity of polymeric filaments is known less than 1 ($em < 1$) due to the filament fineness. Thus, it is necessary to find the relationship between the emissivity of filament and the filament diameter.

In that regard, the two heated, fixed and polished drums with two different diameters 100 mm (on the off-line drawing device, section 3.2.3) and 193 mm (on the melt spinning machine, section 3.2.2) (Figure 3.10) were employed. This method has been successfully utilized by our research group with following assumptions [176, 190, 191]: no or small emissivity of the polished drums, heated drum and filament temperature are equal.

The spun PLA/PVAL monofilament (filament) was wrapped around the polished drums, which were electrically heated (Figure 3.10). The drum and the filament temperatures were measured using the infrared camera at different temperatures over the ranges from 50 °C to 200 °C and different filament diameters ranging from ca. 140 to 380 µm. Figure 3.11 plots the emissivity of the thick PLA/PVAL 30/70 blend filaments versus their diameter in comparison with the emissivity of the fine Polyetheretherketone (PEEK) filament [190]. It is seen that the emissivity of PLA/PVAL increases with the increasing of filament diameter and nearly reaches a constant value of ca. 0.8 with filament diameters more than 300 µm. The emissivity value of the present study for the thick PLA/PVAL filament is around 0.8 much more than that of the investigation by Golzar et al. [190], when they studied the dependence of the emissivity of fine Polyetheretherketone (PEEK) filament on its diameter. However, it is seen

Experimental

that the emissivity profiles versus diameter of both studies seem to be have a good relation and have the similar increasing tendency with the increase of the filament diameter.

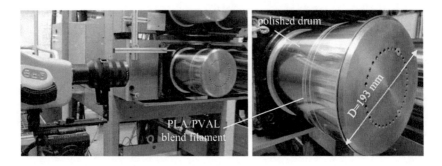

Figure 3.10 Photographic of test stand for measurement the emissivity as correction factor

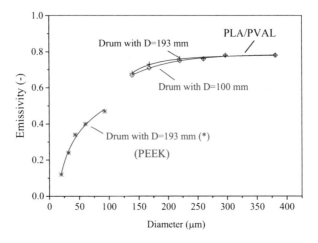

Figure 3.11 The emissivity vs. the diameter of PLA/PVA blends in comparison with pure PEEK filament (*) adapted from Ref. [190]

Figure 3.12a shows a typical image obtained from a snapshot of filament temperature measurement using infrared camera. The temperatures are performed as digits in every pixel in the image. Figure 3.12b present the three different temperature profiles (L1, L2, and L3) perpendicular to spinning line (filament). The maximum value of each temperature profile is assumed to be the uncorrected temperature of the filament. The filament temperatures were then corrected by the emissivity, which is depended on the filament diameter as presented in Figure 3.11.

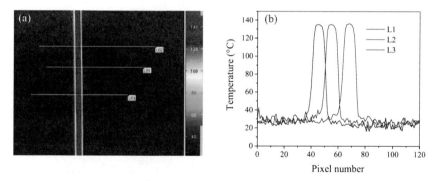

Figure 3.12 A snapshot of filament temperature measurement using the infrared camera (a) and the three temperature profiles perpendicular to spinline (b)

3.3.7 On-line tension measurement

Measurements of filament tension were carried out using a Tensiometer DIGITENS® 485 for hand-held measurement with the measuring head Type 125.120.1 by Honigmann Industrielle Elektronik GmbH. Before the measurement of the filament tension, the Tensiometer was set at zero and was calibrated using various metal plates of 2, 5, and 10 grams, which attached to the PLA/PVAL blend filament. During the blend filament was manually moved up and down through the measuring head, the full range of the Tensiometer was adjusted.

For the tension measurement of the running filament, the measuring head was located at 180 cm from the die exit just above the take-up device. Ten measurements were manually observed and an average value was calculated.

3.3.8 Fiber-capturing device

Pieces of PLA/PVAL blend filaments 4 cm long were collected using a self-constructed fiber-capturing device which was fabricated in our own machine shop at IPF Dresden (Figure 3.13). The device is mounted on a platform that can be moved vertically over distances ranging from 2 cm to 150 cm from the die exit to capture the running filament at different locations along the spinline. The fiber capturing device consists of several changeable clamps and it is automatically operated by compressed air. The molten polymer filament is caught very fast within 0.01 s and is instantly quenched and solidified as soon as it was trapped by surrounding air at room temperature of 25 °C without any additional cooling medium. The solidified pieces of PLA/PVAL blend filaments between the clamps are then ready to measure their diameter and to investigate their morphological properties.

A similar fiber capturing device has already been used to cut a specific fiber length for calculation the linear density of fiber by Kase and Matsuo [153] and determination the fiber diameter by Ishibashi et al. [192], and Oh [193].

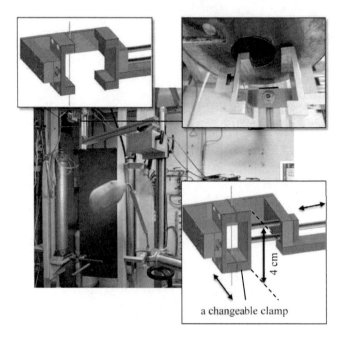

Figure 3.13 Fiber-capturing device and schematic view of a captured molten filament

3.3.9 Filament diameter measurement via light microscopy

The filament diameters both in cross-section (Figure 3.14) and longitudinal direction (Figure 3.15) versus distance x from the die exit were measured using a Keyence digital microscopy VH-Z100 zoom lens with 100 – 1000×magnification. For the filament diameter measurement in the cross-section and in the longitudinal direction, 10×5 diameters and 2×5 diameters, respectively, at each position were manually measured. For the measurement the filament diameter in the longitudinal direction, a PEAK glass (GWJ Company, US) with the scale length of 50 mm and scale division of 0.1 mm (Figure 3.15) was used to fix the captured filament and measure at right position, which is considered as the distance x from the die exit along the spinline.

Figure 3.14 Measurement of the filament diameter in the cross-section of PLA/PVAL filaments using microscopy (left) and the photograph of filament keeping device (right)

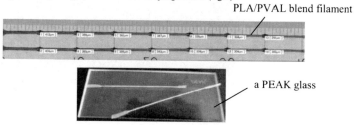

Figure 3.15 Measurement of the filament diameter in the longitudinal direction (above) and the photograph of a PEAK glass and the two captured filaments obtained using fiber capturing device (below)

3.3.10 Mechanical characterization

Mechanical characterization of PLA/PVAL blend filaments

The tensile and elongation properties of the filaments were determined by tensile testing using a Zwick/Roell Z 0.5 Universal Tester controlled with TestXpert software (Zwick/Roell, Ulm, Germany). The gauge length was 100 mm and the test speed was 200 mm·min^{-1}. Twenty four tests were performed for 24 filaments. The fineness and diameter were measured by weighing and using an optical microscope, respectively. The measurements were done in an air-conditioned laboratory with 22 °C and 65 % relative humidity.

Mechanical characterization of nanofibrillar PLA scaffolds

The tensile strength and elongation of nanofibrillar PLA scaffolds were carried out using a Zwick/Roell Z 2.5 Universal Tester controlled with TestXpert software (Zwick/Roell, Ulm,

Experimental

Germany) according to ASTM Standard No. D882[1] with the gauge length of 50 mm and the test speed of 5 mm·min^{-1}.

The preparation of the PLA scaffolds is shown in Figure 3.16. For each scaffold, 5×24 off-line drawn PLA/PVAL filaments (DR=1.5) were wound and fixed in the brass frame during the moving process to ensure the width of scaffolds is the same value of 10 mm width. The thickness of the scaffolds was measured using a Mitutoyo digital micrometer with 0.001 mm discrimination. For each scaffold, the thicknesses of 5×3 different positions (5 positions in the middle and 5×2 in the both sides) along a scaffold (3 blue lines) were measured (Figure 3.17). The mean thickness of each scaffold is used for calculation the tensile stress. In the present study, fifteen scaffolds were prepared for determining their mechanical properties.

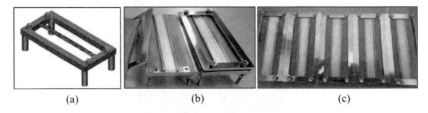

(a)   (b)   (c)

Figure 3.16 Preparation of PLA scaffolds for the tensile testing: (a) self-constructed frame, (b) off-line drawn PLA/PVAL filaments wound and fixed in the frame, (c) the filaments are then immersed in water during removing process

Figure 3.17 Thickness measurement of a PLA scaffold

---

[1] Originally, ASTM–D882 is the standard test method for tensile properties of thin plastic sheeting and film less than 1 mm

3.3.11 Morphology characterization

Several kinds of PLA/PVAL blend samples were prepared to study their morphology: the PLA/PVAL blend granulates, PLA/PVAL blend filaments taken from the bobbin, the PLA/PVAL blend fragments 1 cm long cut from the middle of the captured PLA/PVAL blend filament 4 cm long, and the PLA/PVAL blend fragments fractured at the middle of the captured PLA/PVAL filament 4 cm long in liquid nitrogen. The last one was prepared to investigate the cross-sectional morphology of PLA/PVAL blend filaments.

The blend samples mentioned above are immersed in chloroform for 8 hours at 50 °C or/and in distilled water for 24 hours at room temperature (ca. 25 °C) to remove dispersed PLA phase or/and PVAL matrix material, respectively. In the latter case, the remaining dispersed PLA phase after removing the PVAL matrix is unstable during removing process. Therefore, the self-fabricated filament keeping device was used to fix the PLA/PVAL blend samples, which are laid on flat filter paper or filter stainless metal during removing PVAL in water for 24 hours (Figure 3.18).

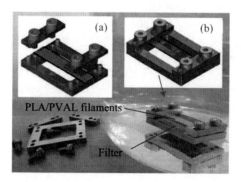

Figure 3.18 PVAL removing process in distiller water: samples were fixed in filament-keeping device (a and b), then were immersed in water for 24 hours.

Scanning electron microscopy (SEM)

After etching dispersed PLA phase or removing PVAL matrix from PLA/PVAL blend samples, the remaining phase was dried at room temperature for 24 hours. All dried samples were investigated using Scanning Electron Microscopy (SEM) Ultra plus (Carl Zeiss NTS GmbH, Oberkochen, Germany). The sample discs were prepared by sputtering a thin layer of 3 nm platin.

Experimental

Image analysis

SEM images of PLA/PVAL cross-section can be easily evaluated using Scandium Image Analysis Software (SIA) (Olympus soft imaging solutions GmbH, Münster, Germany). This program automatically detects the dispersed PLA phase (black holes) (Figure 3.19). The size and shape of dispersed PLA domains such as perimeter, circular equivalent diameter (CED), aspect ratio, circularity, etc. can be easily obtained using this program. However, the program can only be helpful if the holes and background of the images have the considerable differential contrast. Therefore, in this study, the shape and size of dispersed phase were measured both manually and automatically with the SIA program. The CED d_{CED}, the number-average CED $\bar{d}_{CED}$, and circularity/form factor were used to calculate and evaluate dispersed PLA phase domains. These parameters are defined as follows (Figure 3.20) [194, 195]:

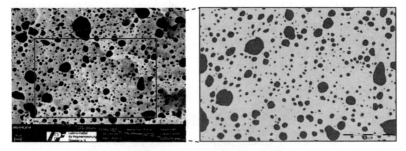

Figure 3.19 An original SEM image of PLA/PVAL blend (a) and the image analyzed using SIA software (b): The blue domains and grey background represents the dispersed PLA phase and PVAL matrix, respectively.

Circular equivalent diameter (CED) or equivalent area diameter, d_{CED}: Diameter of a circle having the same area as the particle. It can be calculated using the Equation 3.4

$$d_{CED} = \sqrt{\frac{4A}{\pi}} \qquad (3.4)$$

where A is the area of the particle

Number-averaged CED $\bar{d}_{CED}$ was defined as follows:

$$\bar{d}_{CED} = \frac{\sum n_i d_{CEDi}}{\sum n_i} \qquad (3.5)$$

where n_i is the number of dispersed phase domains with d_{CED}

Circularity or form factor, C_n: There are several definitions of the circularity [196]. In this study, the circularity is evaluated according Cox's equation (Equation 3.6), where P is perimeter. It is the ratio of the area of the particle to the area of a circle with the same perimeter. $C_n = 1$ indicates a perfect circle and $C_n \to 0$ points out to increasingly elongated polygon.

$$C_n = \frac{4\pi A}{P^2} \tag{3.6}$$

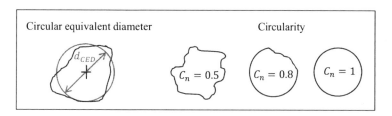

Figure 3.20 Illustrating of circular equivalent diameter (left) and circularity of a 2D particles (right)

The size/diameter of the dispersed PLA phase in PLA/PVAL blend filaments after removing the PVAL matrix was also manually measured using SIA. For the ellipsoidal droplets, the maximum sizes are selected to determine their diameters (Figure 3.21).

Figure 3.21 An example of the measurement of dispersed PLA phase determined using SIA

Chapter 4
Results and discussion

4.1 Miscibility of PLA/PVAL blends

As discussed in chapter 2, section 2.2, the phase morphology of polymer blends that results in micro-and nanofibrillar structures depends on material parameters and processing conditions (shear stress, shear rate, composition, viscosity ratio, interfacial tension, etc.). The viscosity is known to be one of the most critical variables for controlling blend morphology of immiscible polymer blends. Generally, if the minor component has a lower viscosity than the major one, the minor component will be finely dispersed in the major one. On the contrary, the minor component will be coarsely dispersed if its viscosity is higher than viscosity of the major one [44, 46, 126, 197]. At first we consider the viscosity of neat PLA, PVAL used in this study. Based on the rheological properties of these two polymers, for instance Melt Flow Index (MFI) and zero-shear viscosity (chapter 3, section 3.1), it can be clearly seen that the viscosity of PVAL at same conditions is much higher than that of PLA. Therefore, the PLA will form a dispersed phase which further deform into micro-nanofibrils in the PVAL matrix only if the PLA and PVAL are the minor and major component, respectively. To limit the number of blend components, therefore, PLA/PVAL blend components with dispersed PLA phase less than 50 percent (by weight) was investigated.

Micro- and nanofibrillar structures from polymer blends only occurs for immiscible polymer blends, i.e. multiphase polymer blends. Thus, the basic requirement for producing the nanofibrillar structures must be satisfied is that polymer blends are thermodynamically immiscible under certain conditions [198].

Various methods have been used to determine the miscibility of polymer blends such as film clarity, glass transition temperature, microscopy, etc. Each technique has its own advantages, sensitivity and an application for a particular polymer system. In this study, the considerations of glass transition temperature(s) T_g obtained by using DSC were used to determine the miscibility of polymer blends. In principle, immiscible polymer blends are phase separated and exhibit glass transition temperature of each component. On the contrary, a single glass transition temperature is often taken as evidence of the formation of a miscible blend. The miscibility of PLA and PVAL blends were also studied by using the Fourier transform infrared (FTIR) spectra. In principle, if there are no specific interactions between the two polymers, the two polymers are immiscible.

Phase separation morphology of polymer blends visually observed using Scanning Electron Microscopy (SEM) is also considered as a criterion for miscibility of polymer blends. For immiscible polymer blends, each phase in the blends is clearly detected under SEM and Atomic Force Microscopy (AFM) independent of the blend ratio. For the SEM investigations, the solvent extraction often be used to remove the selected phase and the remaining holes are observed.

4.1.1 Glass transition temperature

Table 4.1 summarizes the glass transition temperatures ($T_g s$) of neat PLA, PVAL, and PLA/PVAL blends with different blend ratios by weight percent. Figure 4.1 shows the thermograms obtained in second heating run for neat PLA, PVAL and their blends.

Neat PLA shows two melting peaks at around 160 °C and 167 °C. Neat PVAL exhibits a very broad melting peak at around 174 °C (Figure 4.1). These results indicate that the both PLA and PVAL are semi-crystalline polymers.

Neat PLA, PVAL exhibits a glass transition temperature around 61 °C and 34 °C, respectively (Figure 4.1). Due to the difference of $T_g s$ between neat PLA and neat PVAL, the miscibility of polymer blends can be studied by the measuring the $T_g s$ of the two components. It is well known established that immiscible polymer blends clearly demonstrate two T_g values for respective neat components that are independent of composition. On the contrary, the polymer blend is likely to be miscible if only one T_g is observed. For almost blend ratios of PLA /PVAL, except the low content of PLA (10 % PLA), two T_g are found. However, the T_g of each component slightly shifts to the T_g of another component with the increase of another component content (Figure 4.2). This suggests that PLA/PVAL blends are immiscible

Results and discussion

polymer blends in amorphous region, but there are some interactions between the two components that have an influence on the phase behavior of these binary polymer blends.

Table 4.1 The glass transition temperature of PLA, PVAL, and their blends

Blend ratio of PLA/PVAL by weight percent [wt%/wt%]	Weight [mg]	$T_{g,PLA(*)}^{2nd\ heating}$ [°C]	$T_{g,PVAL(*)}^{2nd\ heating}$ [°C]
100/0	5,261	61	-
50/50	5,403	57	44
40/60	5,582	59	44
30/70	5,541	55	41
20/80	5,100	55	40
10/90	5,924	56	37
0/100	5,638	-	34

(*) determined using inflection point method

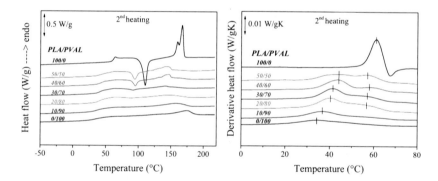

Figure 4.1 DSC thermograms obtained from the 2nd heating runs of neat PLA, PVAL, and PLA/PVAL blends: (a) heat flow vs. temperature, (b) derivative heat flow vs. temperature in the glass transition range

The hydroxyl groups in PVAL can form hydrogen-bonds with carbonyl (C = O) groups of various polyesters [61, 110, 198, 199]. However, most of studies have found that the PLA/PVAL system is immiscible or partially miscible [61, 200, 201]. In this study, the PVAL hydrolyzed 88 % was prepared by the saponification of poly(vinyl acetate) (PVAc) having vinyl alcohol and vinyl acetate groups [202]. In other words, the chemical structure of PVAL includes both the hydroxyl groups (-OH) and carbonyl (-OCOCH3), as shown in Figure 4.3, is called poly(vinyl alcohol-*co*-vinyl acetate). In this case, the formation of hydrogen bond in PLA/PVAL blends becomes more complex because PVAL itself may form inter- and intramolecular hydrogen bond.

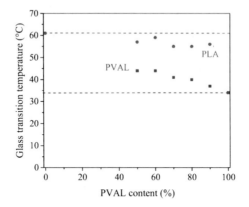

Figure 4.2 The dependence of glass transition temperature T_g on the blend ratios

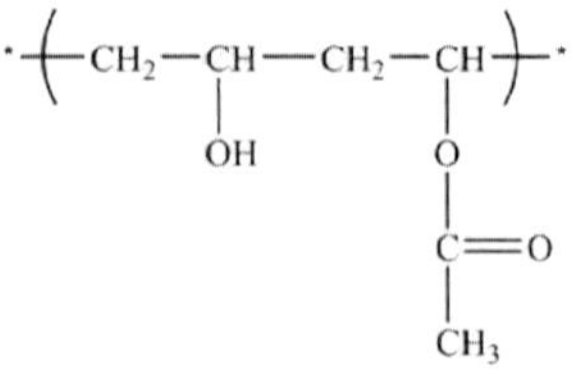

Figure 4.3 Chemical structure of poly (vinyl alcohol-*co*-vinyl acetate)

4.1.2 ATR-FTIR spectroscopy

ATR-FTIR has been used as a useful and powerful tool for the investigation of specific functional groups, chemical bonds that exist in a material, or chemical interactions between two polymers. In principle, if there are specific chemical interactions between the two polymers, the ATR-FTIR spectra of polymer blends either exhibit obvious and significant shifts of the existing bands or new bands appear.

Figure 4.4 shows the ATR-FTIR spectra of PVAL and PLA. Table 4.2 summarizes their major band assignments. The spectrum of PVAL shows the band at 3264 cm^{-1} between 3550 and 3200 cm^{-1} related to the O $-$ H stretching vibration as well as to vibrations from the intermolecular and intramolecular bonded hydroxyl groups. The observed bands between

3000 and 2840 cm^{-1} represent the stretching vibrations of methylene from alkyl segments of PVAL. The medium band between 1750 and 1700 cm^{-1} related to the stretching vibration of carbonyl C=O group (the polyvinyl acetate fraction). The peaks at 1241, 1089 and 1039 cm^{-1} are due to the stretching vibrations of C − O − C and C − OH groups [203-207]. This result confirmed that the PVAL Moviflex TC 232 includes acetate groups in polymer chain. The spectrum of PLA shows the peaks at 2996, 2946 and 2881 cm^{-1} assigned to the stretching vibrations of $CH_{3(asymmetric)}$, $CH_{3(symmetric)}$, and $CH_{2(symmetric)}$ groups, respectively. The band at 1746, 1180, and 1080 cm^{-1} are signed to the stretching vibration of C = O, C − O, and C − O − C of ester group, respectively [90, 208-210].

Taking into account the functionalities of PVAL and PLA confirmed spectroscopically, there may be the existence of two types of hydrogen bond interactions: (1) hydroxyl-carbonyl hydrogen-bonds due to inter- and intramolecular interactions between the hydroxyl and carbonyl groups of PVAL and between hydroxyl groups of PVAL and carbonyl groups of PLA; (2) hydroxyl-hydroxyl hydrogen-bonds may be also formed by interactions of adjacent hydroxyl groups and hydroxyl-carbonyl groups (Figure 4.5). Therefore, it is difficult to detect the hydrogen bond between PVAL and PLA from the hydroxyl bands [61].

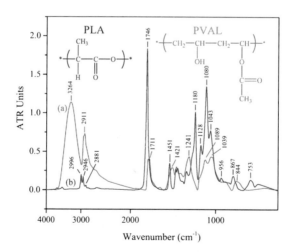

Figure 4.4 ATR-FTIR spectra of PVAL (a) and PLA (b)

Results and discussion

Table 4.2 Main ATR-FTIR vibrational bands in PVAL (88 % hydrolyzed) and in PLA

Material	Wavenumber (cm^{-1})	Assignment
PVAL	3264	O—H stretching and adsorbed water
PVAL	2911	C—H stretching
PVAL	1711	C=O stretching (acetate groups)
PVAL	1089 and 1039	C—O—C and C—OH stretching
PVAL	1421	CH$_2$ bending
PVAL	1241	C—O (crystallinity)
PVAL	844	C—C
PLA	2996 and 2946	C—H stretching from alkyl groups
PLA	2881	C—H stretching from CH modes
PLA	1746	C=O stretching
PLA	1451	CH$_3$
PLA	1180 and 1080	C—O—C stretching

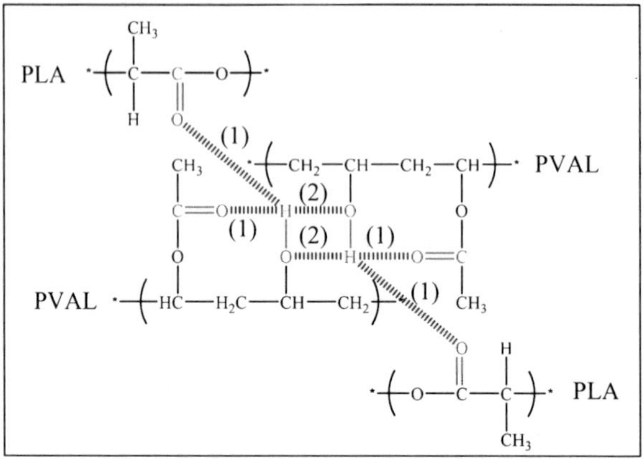

Figure 4.5 Various potential hydroxyl- hydroxyl (1) and hydroxyl-carbonyl (2) hydrogen-bonds between PVAL and PLA

Figure 4.6 shows the comparison of the ATR-FTIR spectra of PVAL and PLA with PLA/PVAL blends. In the hydroxyl stretching region between 3550 – 3200 cm^{-1}, it can be obviously seen that as amounts of PVAL increase in blends, the hydroxyl band becomes more intensive, that means an increase of the hydroxyl group concentration, but there is almost no band shift of this group. In the carbonyl stretching zone (1750 – 1710 cm^{-1}), the carbonyl band becomes more intensive with increasing the PLA content, which is related to the increase of the carbonyl group fraction of PLA and the decrease of the carbonyl groups of PVAL. This tendency shows that the amount of carbonyl group of PLA is much higher than the amount of carbonyl group in PVAL. Shuai et al. [61] reported that this strong carbonyl absorption of PLA could be more indicative to detect hydrogen bond exiting in PLA/PVAL blends. In principle, if there is quite strong formation of hydrogen bonds, the carbonyl absorption should shift to lower frequency (or longer wavelengths) [211].

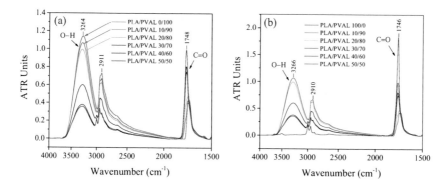

Figure 4.6 ATR-FTIR spectra comparison of PVAL/PLA blends with PVAL (a) and with PLA (b) over the range of 4000 – 1500 cm^{-1}

To confirm the shift of carbonyl band positions precisely, the second derivatives of ATR-FTIR spectra of both blends and neat PVAL, PLA were analyzed. Figure 4.7 shows both the ATR-FTIR spectra and their second derivatives of the PVAL, PLA, and PLA/PVAL blends in the carbonyl region (1800 – 1650 cm^{-1}). It can be seen that there is almost no shift of the carbonyl peak in the ATR-FTIR spectra. Even in the second derivative of spectra, only a slight shift to lower frequencies was found with the increase of PLA content. This may be due to the formation of intermolecular hydrogen bonds between hydroxyl and carbonyl groups of the two chains of different polymers. But, on the other hand the band position is moving to a usual carbonyl position of PLA at 1746 cm^{-1} (Figure 4.7b, bottom), what is caused by the differences in crystalline structures [212]. This observation agreed well with the T_g values obtained by the DSC: the T_g of PLA in PLA/PVAL blends shift to higher temperatures with the increase of PLA content (Figure 4.2). Simultaneously, carbonyl bands positions of PVAL don't exhibit any changes (Figure 4.7a, bottom). As a result, it is impossible to distinguish the PLA/PVAL fraction of hydrogen bonding interaction in such blend system unambiguously.

Results and discussion

As a summary, one can conclude that PVAL and PLA are immiscible or partial miscible polymer blends.

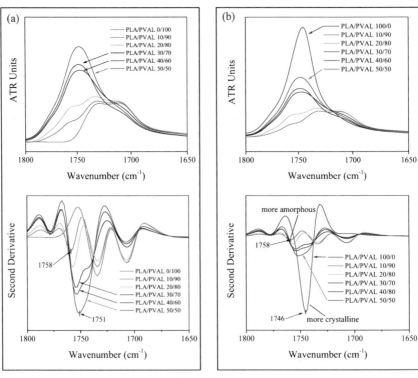

Figure 4.7 The ATR-FTIR spectra (top) and the second derivatives of the ATR-FTIR spectra (bottom) of the PVAL, PLA and their blends in the carbonyl region: (a) PVAL, PLA/PVAL blends, (b) PLA, PLA/PVAL blends

4.1.3 SEM and atomic force microscopy (AFM) images

The phase morphologies of fractured PLA/PVAL blends at different blend ratios in weight percent (10/90, 20/80, 30/70, 40/60, 50/50 wt/wt %) were performed using SEM imaging technique (chapter 3, section 3.3.11). Figure 4.8 shows SEM images of fracture surfaces of PLA/PVAL blends after etching PLA phase. The SEM images indicate that PLA/PVAL blends are immiscible over the entire blend ratios from 10/90 to 40/60. The dispersed PLA phase becomes more uniform in both size and shape with increasing PVAL content.

Results and discussion

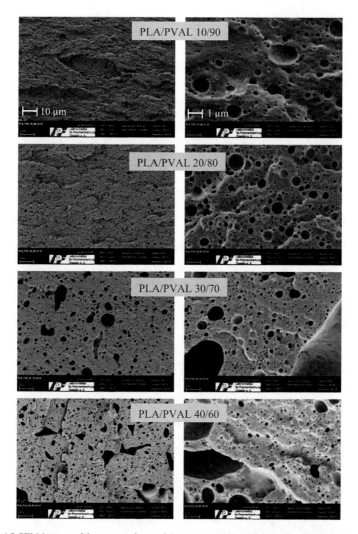

Figure 4.8 SEM images of fracture surfaces of the as-extruded PLA/PVAL blend after etching of the PLA phase at various PLA/PVAL blend ratios: scale bars for the left and right column are 10 µm and 1 µm, respectively

In principle, for a completely immiscible polymer blend, the morphology could be observed clearly under SEM or even by optical microscopy without removing any phase [53]. Figure 4.9 shows the SEM images of a microtome surface of PLA/PVAL 30/70 blend without etching PLA phase. It is seen that the PLA phase is well dispersed in the PVAL matrix, whereas PLA domains have a convex surface. However, the boundary between the PLA domains and PVAL matrix is not very sharp (Figure 4.9b). Thus, it allows one to confirm again that PLA/PVAL blend is an immiscible system with the existence of some specific interaction between the two polymers.

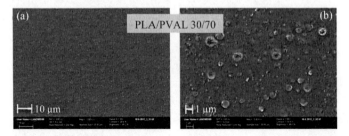

Figure 4.9 SEM mages of a microtome surface of PLA/PVAL 30/70 blend without etching the PLA phase: (a) scale bar: 10 μm, (b) scale bar: 1 μm

Figure 4.10 represents the Atomic Force Microscopy (AFM) images of a microtome surface of PLA/PVAL 30/70 blend obtained using a Dimension ICON (Bruker-Nano, USA) in the peak force tapping mode. The silicon nitride sensors SCANASYST-AIR (Bruker, USA) with a nominal spring constant of 0.4 $N \cdot m^{-1}$, the tip radius is 2 nm, was used. Similar with the SEM images, in the AFM images the PLA domains were well dispersed in PVAL matrix and appeared brighter than PVAL the matrix. The boundaries between the PLA domains and PVAL matrix are not also seen very clear. Thus, one can conclude again that the used PLA/PVAL polymers are immiscible with each other, but there is existence of specific interaction between them.

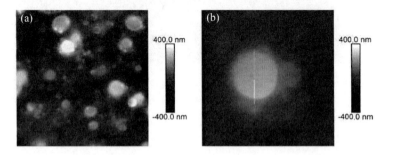

Figure 4.10 AMF images of a microtome surface of PLA/PVAL 30/70 blend: (a) scale bar: 4 μm; (b) scale bar: 800 nm.

4.2 Melt Spinning of nanofibrillar structures from PLA/PVAL blends[1]

This section presents a novel fabrication process for preparing continuous thermoplastic biodegradable nanofibrillar PLA structures from PLA/PVAL by using the conventional melt spinning method on the industrial-scale melt spinning device with twin-screw extruder (section 3.2.2). PLA, PVAL, and their blends in different blend ratios were extruded in a corotating twin-screw microcompounder (section 3.2.1). The resultant extrudates were then characterized their microfibrillar and lamellar hybrid morphology in order to fix the PLA/PVAL blend ratio for the melt spinning process. The thermal and rheological properties of the blend extrudates were also analyzed to select an optimal set of spinning process parameters. After melt spinning, the textile-physical properties of resulting PLA/PVAL blend filaments were investigated. For characterization of nanofibrillar structures, PLA/PVAL blend filaments were immersed in distilled water to remove completely PVAL matrix and then dried at room temperature. The dried PLA nanostructures obtained were examined using a scanning electron microscope (SEM).

4.2.1 Blend ratio and spinning temperature profiles for the melt spinning process

Determination of blend ratio based on SEM images

Before doing any melt spinning process of polymer blend that results into micro-/nanofibrillar structures, a lot of parameters such as blend ratio, processing temperatures, etc. should be firstly determined. This pre-experimental step is often done on microcompounder. Figure 4.11 shows an experimental setup to get the blend samples for the morphological study. There are two types of blend samples: the blend extrudates without stretching and the blend filament taken from the bobbin.

Figure 4.12 presents SEM images of PLA structures from PLA/PVAL blend extrudates (the left column) and PLA/PVAL blend filaments (the right column) after removing the PVAL matrix produced from different blend ratios ranging from 90/10 to 60/40 wt%/wt%. The PLA structures from PLA/PVAL 50/50 are not shown in Figure 4.12, because the PLA/PVAL 50/50 blend has not successfully processed.

It can be seen that while in PLA/PVAL blend filaments, continuous PLA fibrillar structures existed in all blends, they did not appear in blend extrudates when the amount of PLA in the

---

[1] Parts of this section have been published in: N.H.A. Tran, H. Brünig, C. Hinüber, and G. Heinrich. *Macromol. Mater. Eng.* 299(2), 219-227 **(2014)**.

Results and discussion

blends was 10 wt% (Figure 4.12a). In blend filaments, the PLA fibrillar structures became more uniform and smaller in size with increasing amounts of PVAL in blends. However, by blending a higher amount of PVAL (80 & 90 wt%) with PLA, the remaining amount of PLA fibrillary structures obtained were quite a few (Figure 4.13). In this study, therefore, the PLA/PVAL 30/70 blend ratio was selected to produce PLA nanofibers. In principle, one can select any of blend ratios as mentioned above with PLA content less than 60 % to get nanofibrillar structures. Another requirement must be satisfied is that the spinnability of PLA/PVAL blends. This is discussed in the following section.

Figure 4.11 Experimental setup on microcompounder (adapted from the oral presentation at the 30th international conference of the Polymer Processing Society, 08-12 June, 2014, Cleveland, Ohio, USA)

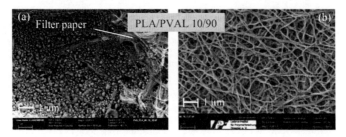

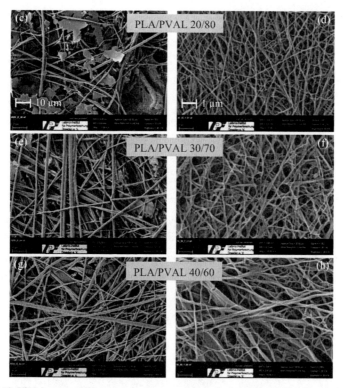

Figure 4.12 SEM images of PLA structures prepared from the as-extruded PLA/PVAL blends that were obtained using the microcompounder (left column, scale bar: 10 μm, except Figure 4.12a with scale bar 1 μm), and from the PLA/PVAL blend filaments (right column, scale bar: 1 μm)

Figure 4.13 Photographs of blend filaments before (above) and after (below) removing PVAL matrix

Results and discussion

Determination of spinning temperature profiles

Most melt-spinnable polymers are temperature sensitive and degrade rapidly at high temperatures. But, the lower the spinning temperature the higher the pressure required to force the molten polymer through the spinneret. Non-uniform flow, and melt fracture can occur if the spinning temperature is too low. Consequently, a compromise is always made between a high spinning temperature for reasonable spinning and a low spinning temperature for a minimum of polymer degradation [37]. Figure 4.14 shows photographs of molten PLA/PVAL 30/70 blend taken after two hours of a trial spinning process. Thermal decomposition of this polymer blend increases with residence time, especially under the simultaneous effect of high pressure. In practice, PVAL has been processed mainly from aqueous solutions [213-217]. The melt spinning of PVAL has not been succeeded industrially in spite of many efforts, because of its thermal degradation and high viscosity properties [38, 39, 218, 219]. Thermal behaviour of PLA/PVAL blends is totally dominated by thermal properties of PVAL during melt spinning process. Therefore, it is necessary to find the appropriate melting temperature profile for melt spinning process of PLA/PVAL 30/70 blend. For this purpose, thermal behaviours and rheological properties of neat polymers (PLA, PVAL) and their blends were investigated.

Figure 4.14 Photographs of molten PLA/PVAL 30/70 in the spinning head after two hours of an unsuccessful trial spinning process

Thermogravimetric Analysis (TGA)

Figure 4.15 shows the results of non-isothermal TGA measurement of neat polymers (PLA, PVAL) and their blends. There are two distinct and well-separated turns in thermogravimetric (TG) curves (Figure 4.15a) and two corresponding mass-loss peaks in derivative thermogravimetric curves (DTG) (Figure 4.15b) for neat PVAL and PLA/PVAL blends. Thus, the thermal degradation of neat PVAL and PLA/PVAL blends can be roughly regarded as two-step-degradation. While PLA was not degraded below temperatures of 260 °C, PVAL started to decompose just above the melting point T_m ($T_{m,PVAL}$=175 °C), and the decomposition temperature T_d (temperature at the 5% mass loss) is 190 °C. The melting point and the decomposition temperature of PVAL are close to each other due to the multi-hydroxyl structure. The thermal degradation also undergoes simultaneously during melting. This makes it difficult to continue the spinning process stable for a long period of time. By means of using isothermal study, the thermal decomposition of materials and its time dependence can be evaluated. In the isothermal method, decomposition time is estimated for a defined range of mass loss. From Figure 4.16, it was found that the thermal degradation of neat PVAL and their blends with PLA increases with the resident time. By blending with an increasing amount of PVAL in blends, the thermal degradation blends was more stable. The minimum thermal degradation of blends was found when amount of PVAL in blends was varied from 70 to 90 wt%. The thermal degradation of PLA/PVAL 50/50 blend increased rapidly after 20 min of heating at 200 °C. The PLA/PVAL 50/50 blend lost 25.2 % of its original mass 50 min of heating at 200 °C (Figure 4.16).

The dependence of the mass loss, according to isothermal temperature of PLA/PVAL 30/70 blend is shown in Figure 4.17. Isothermal TG curves were obtained by heating the sample at different temperatures of 185, 190, 200, and 205 °C. It is evident that the higher temperature, the lower time necessary for the same mass loss to occur. At a temperature of 205 °C, the mass loss of PLA/PVAL blend occurred more early and rapidly than that at lower temperatures. Therefore, it is suggested that the PLA/PVAL blend should be processed at a maximum temperature of 200 °C.

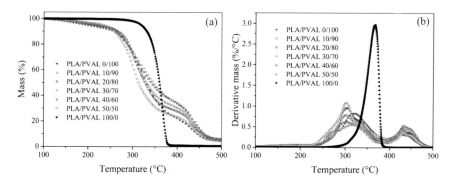

Figure 4.15 Mass vs. temperature (a) and derivative mass vs. temperature (b) of PLA, PVAL, and their blends in TG measurement

Results and discussion

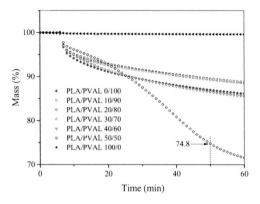

Figure 4.16 Mass vs. time in isothermal TG measurement of PLA, PVAL, and their blends at 200 °C

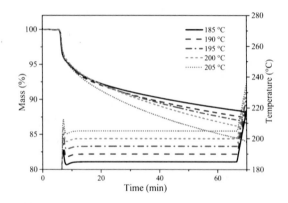

Figure 4.17 Mass vs. time in isothermal TG measurement of PLA/PVAL 30/70 blend at temperatures of 185, 190, 195, 200, and 205 °C

Rheological behavior of PLA, PVAL, and their blends

In the range of shear rates investigated, the complex viscosity of the molten PLA, PVAL and their blends decreases with increasing shear rate, showing a typical property of non-Newtonian fluids and all follow the shear-thinning behaviour (Figure 4.18). All curves were fitted well to a Carreau-Yasuda fluids model [220, 221]. It is worth noting here that the viscosity of PVAL as well as PLA/PVAL blends is extremely high in comparison with the viscosity of typical spinning polymers, like PLA. From Figure 4.18b, it can be seen that there is a huge difference of the viscosity levels between the two polymers. This viscosity difference between PLA and PVAL is also known one of the key conditions that must be satisfied to form the less viscous PLA dispersed phase into micro-/nano fibrillar structures in the more viscous PVAL matrix [126].

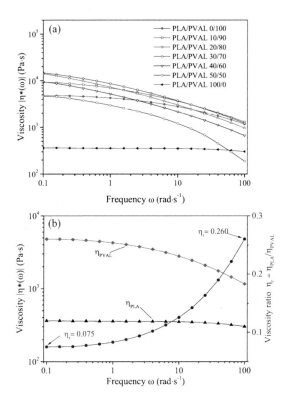

Figure 4.18 Complex viscosity vs. frequency for PLA, PVAL, and PLA/PVAL blends (a) and viscosity ratio of PLA/PVAL (b) at 195 °C

A large variation in shear viscosity was observed with PLA/PVAL blend ratio of 30/70 at different temperatures. With increasing the temperature, the complex viscosity of PLA/PVAL 30/70 blend decreases (Figure 4.19), it enables for the melt spinning of polymers with high viscosity, like PLA/PVAL easier due to the decrease of pressure during spinning. But, the higher the temperature the more the degradation of polymers will occur. Melt degradation can be determined by recording the dynamic moduli at constant frequency continuously as a function of time at process temperature. PLA/PVAL 30/70 blend is very sensitive to process temperature, as shown in Figure 4.20 for the five different temperatures at constant frequency of 10 rad·s^{-1} in 60 minutes. The effect of time and temperature on the thermal stability is noticeable. The shear viscosity of PLA/PVAL 30/70 blend decreases with the increase of the testing time. However, this tendency was not observed at a testing temperature of 205 °C after 50 minutes of heating. In other words, the viscosity of the PLA/PVAL 30/70 increases again after 50 minutes of heating. This phenomenon may be caused by the decomposition of polymer at high temperature as interpreted in the above section "Themogravimetric Analysis".

In summary, the TGA and rheology experiments confirmed that the PLA/PVAL blend is temperature sensitive and degrades rapidly at high temperatures. The suitable spinning temperature should range from 185 °C to 200 °C.

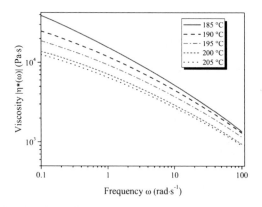

Figure 4.19 Complex viscosity vs. frequency for PLA/PVAL 30/70 blend at different temperatures

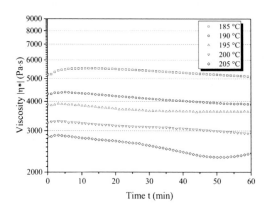

Figure 4.20 PLA/PVAL blend melt stability followed in oscillation at different temperatures at constant frequency of 10 rad·s^{-1}

Results and discussion

4.2.2 Textile-physical properties of PLA/PVAL 30/70 filaments

Effect of spinning velocity

Figure 4.21 demonstrates the tensile and elongation properties of PLA/PVAL blend filaments for different spinning velocities. With increase in spinning velocity from 30 to 50 m·min^{-1} the elongation at break $\varepsilon_R(\%)$ decreases and the tenacity σ_R(cN·tex^{-1}) (the maximum stress) is almost the same. This phenomenon is proving that the increase in spinning velocity increase the molecular orientation in the direction of the fiber axis and therefore enhancing the capability of the fiber to be less elongated. The yield stress σ_y is the highest in the filaments at the spinning velocity of 50 m·min^{-1}, while there is no difference between the filaments at the spinning velocity of 30 m·min^{-1} and 40 m·min^{-1} (Figure 4.21a). However, it is observed from Figure 4.21b that the obtained filaments at the spinning velocity of 30 m·min^{-1} possessed the highest tenacity and elongation at break. Furthermore, it was found that the melt spinning process at the spinning velocity of 30 m·min^{-1} was the most stable process. In this case, therefore, the spinning velocity of 30 m·min^{-1} is the best choice for the melt spinning process of PLA/PVAL 30/70 blend on the industrial-scale melt spinning device.

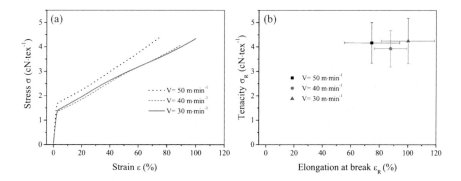

Figure 4.21 Tensile and elongation properties of PLA/PVAL filaments for different spinning velocities: Stress-strain curves (a), tenacity and elongation at break (b).

Effect of off-line draw ratio (DR)

The tenacity of filaments with spinning velocity of 30 m·min^{-1} is about 4.3 cN·tex^{-1}. This value is much less than that of the typical textile fibers. To improve the textile-physical properties of filaments, the obtained filament yarn of 683 tex 24f with spinning velocity of

30 m·min⁻¹ was then off-line drawn with various draw ratios of 1.25, 1.50, 1.60, and 1.70 at a temperature of 75 °C on a laboratory drawing machine at IPF Dresden, which was presented in the section 3.2.3.

Figure 4.22 shows that the drawing process can improve the textile-physical properties of filaments, which, in fact, is well known for polymeric materials. With increasing the off-line draw ratios from DR=1.25 to DR=1.70, the elongation at break of PLA/PVAL filaments decreases while the tenacity increases. These results show a typical physical behavior of textile fibers. Young's modulus E_0(cN·tex⁻¹), tenacity σ_R(cN·tex⁻¹) and elongation at break ε_R(%) that are calculated from these stress-strain curves are listed in Table 4.3.

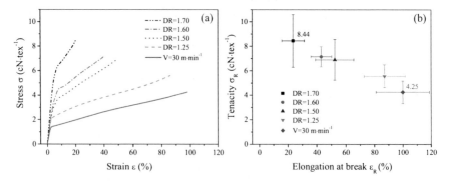

Figure 4.22 Tensile and elongation properties of PLA/PVAL filaments for different draw ratios: Stress-strain curves (a), tenacity and elongation at break (b).

Table 4.3 Textile-physical properties of melt spun/offline-drawn PLA/PVAL filaments

Draw ratio (–)	Young's modulus (cN·tex⁻¹)	Tenacity (cN·tex⁻¹)	Elongation at break (%)
1.00	35.9	4.25	100
1.25	43.8	5.56	87.2
1.50	44.9	6.90	52.4
1.60	47.8	7.15	43.1
1.70	83.8	8.44	23.6

Values of tenacity and Young's modulus of melt spun/off-line drawn PLA/PVAL filaments are plotted as a function of draw ratio in Figure 4.23. As is illustrated in this figure it was found an almost linear relationship between tenacity and draw ratio. Like the tenacity, the Young's modulus initially increases almost linearly with draw ratio from DR=1.25 to DR=1.60. But, at high draw ratio DR=1.70 the modulus tends to approach an upper limit.

The elongation at break of PLA/PVAL blend filaments decreases with increasing draw ratio to reach a value of ε_R =23.6 % for ratios exceeding DR=1.70 (Figure 4.24).

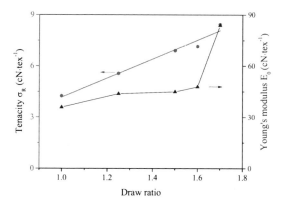

Figure 4.23 Tenacity and Young's modulus versus draw ratio

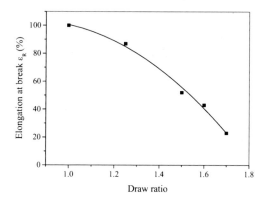

Figure 4.24 Dependence of the elongation at break on the draw ratio

The above analyzed results of textile-physical properties of PLA/PVAL filaments suggest that the optimal spinning velocity for the melt spinning PLA/PVAL 70/30 blend is 30 m·min^{-1} because the filaments exhibit the best tensile and elongation properties. The tenacity of filaments improved significantly from σ_R=4.25 to σ_R=8.44 cN·tex^{-1} by increasing the draw ratio from DR=1.0 to DR=1.70.

Figure 4.25 compares the stress-strain diagram of the PLA/PVAL blend filament with draw ratio DR=1.70 with the stress-strain curves of different commercial textile fibers. The textile-physical properties of these off-line drawn blend filaments with DR=1.70 are similar to that of acetate rayon or wool fibers.

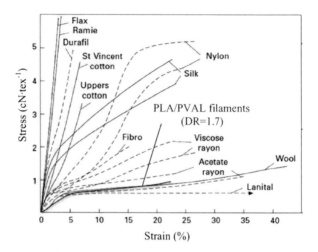

Figure 4.25 Stress-strain curve of PLA/PVAL drawn filaments vs. stress-strain curves of various fibers, modified after [222, 223]

4.2.3 Morphology of nanofibrillar PLA structures from PLA/PVAL filaments

Sample preparation for SEM study

The PLA/PVAL filament yarns[1] (Figure 4.26) produced by using the conventional melt spinning method and the drawn filament yarns obtained by off-line drawing were woven and knitted to produce 2D and 3D textile structures (Figure 4.27). These textile structures were immersed in distilled water to remove PVAL matrix and then dried at room temperature for 24 hours. The dried PLA nanofibrillar structures obtained (Figure 4.28) were examined using SEM.

---

[1] Yarn is a generic term for a continuous strand of textile fibers, filament or material in a form suitable for knitting, weaving, or otherwise intertwining to form a textile fabric (after ASTM Standard D123). In the present study, each yarn is made from 24 PLA/PVAL blend filaments.

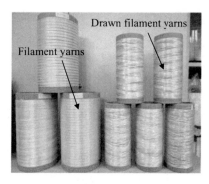

Figure 4.26 Photographic image of PLA/PVAL 30/70 filament yarns at different take-up velocity 30, 40, and 50 m·min^{-1} (3 spools on the left) and off-line drawn filament yarns with different draw ratios (5 spools on the right)

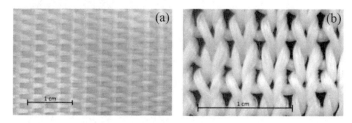

Figure 4.27 Photographic images of woven (a) and knitted fabrics (b) from PLA/PVAL 30/70 blend filaments

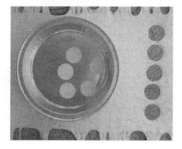

Figure 4.28 Photographic image the dried PLA nanofibrillar structures

Morphology of nanofibrillar PLA Structures

Figure 4.29 shows SEM images of the remaining fibrillar PLA structures obtained from PLA/PVAL blend filaments for different spinning velocities and off-line drawn blend filaments of the PLA/PVAL blend filaments with the spinning velocity of 30 m·min^{-1}. It is

Results and discussion

obviously seen that the dispersed PLA phases form the continuous PLA fibrils in nano scales. The fibril diameter decreases with the increase of spinning velocities as well as the increase of draw ratio. It is worth noting here that the PLA fibrils interconnected together and formed a three-dimensional nanofibrous network. The similar interconnected networks of fibrils are found in several recent studies by Fakirov and co-workers [32, 198, 224]. The formation of the 3-D nanofibrous network may occur due to the potential formation of the hydrogen bonds between PLA and PVAL (Figure 4.5, section 4.1.2). It may also take place during removing process of PVAL matrix or/and during drying process of the remaining PLA phase due to the instability of PLA fibrils.

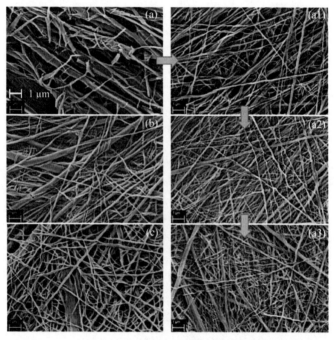

Figure 4.29 SEM images of the remaining PLA nanofibers from PLA/PVAL blend filaments after removing the PVAL matrix: (a) v=30 m·min^{-1}; (b) v=40 m·min^{-1}; (c) v=50 m·min^{-1}; (a1) v=30 m·min^{-1} and DR=1.25; (a2) v=30 m·min^{-1} and DR=1.50; (a3) v=30 m·min^{-1} and DR=1.70; scale bar: 1 μm

The mean diameters of the PLA fibrils are demonstrated in Figure 4.30. It is obviously seen that the mean diameter of fibrils decreases nearly linearly with the increase of spinning velocities. It was also found an almost linear relationship between diameter and draw ratio over a draw ratio range from DR=1.25 to 1.50. At high drawing ratios, the fibril diameter slowly decreases further and seems to be almost constant with a lower limit value of ca. 43 nm corresponding to the draw ratio of 1.70. It is also observed that a few fibrils had been broken-up at the draw ratio of 1.50 (Figure 4.31a). At the higher draw ratio (i.e. DR=1.70), the fibrils no longer deformed and more fractures occurred (Figure 4.31b).

The frequency and cumulative distributions of the diameters of PLA fibrils are also presented in Figure 4.32 and 4.33, respectively. Generally, there are no big different distributions of the diameters of PLA fibrils in all samples. The distributions become smaller with the increase of spinning velocities as well as the increase of draw ratio. It was found that the fibril size distribution of draw ratio DR=1.50 and DR=1.70 is almost identical (Figure 4.32a2, 4.32a3, and Figure 4.33b).

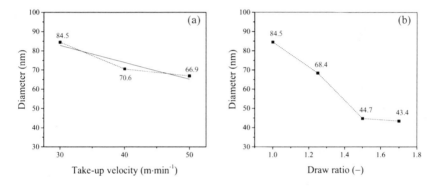

Figure 4.30 Average diameter of PLA fibrils vs. take-up velocity (a) and draw ratio (b)

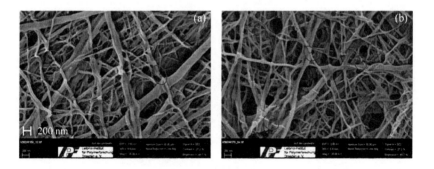

Figure 4.31 PLA fibrils from drawn PLA/PVAL filaments with DR=1.5 (a) and DR=1.7 (b). The fracture of PLA fibrils is marked by a red colour oval. Scale bar: 200 nm

Results and discussion

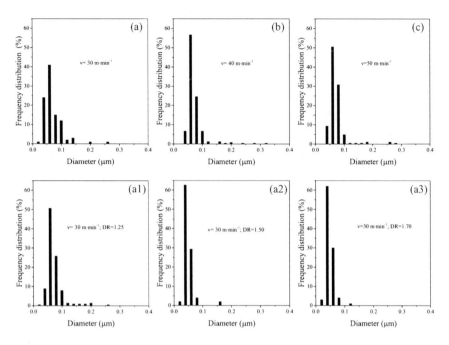

Figure 4.32 Frequency distribution histograms of the diameters of PLA fibrils: (a) v=30 m·min^{-1}; (b) v=40 m·min^{-1}; (c) v=50 m·min^{-1}; (a1) v=30 m·min^{-1} and DR=1.25; (a2) v=30 m·min^{-1} and DR=1.50; (a3) v=30 m·min^{-1}

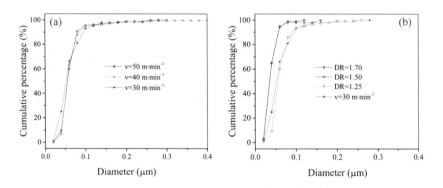

Figure 4.33 Cumulative distributions of the diameters of PLA fibrils for various take-up velocities (a) and various draw ratios (b)

4.2.4 Purity of nanofibrillar PLA structures (PLA scaffolds)

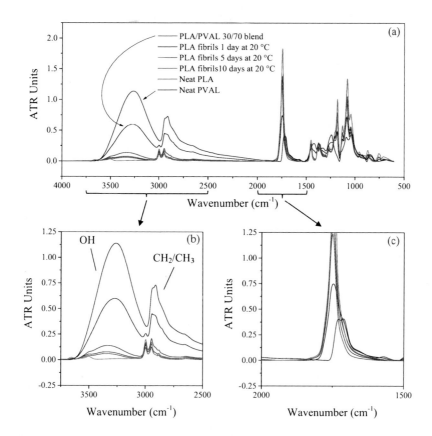

Figure 4.34 ATR-FTIR spectra of neat PLA, PVAL, PLA/PVAL 30/70 blend, and the PLA scaffolds obtained from PLA/PVAL blend filaments, which were immersed in water bath at room temperature ca. 25 °C for 1, 5, and 10 days.

Figure 4.34 and 4.35 show the ATR-FTIR spectra of neat PLA, neat PVAL, PLA/PVAL 30/70 blend, and various PLA scaffolds for different removing conditions. It can be seen from Figure 4.34 that the ATR-FTIR spectra still has both the shoulder of CH2/CH3 vibrations and the O — H stretching vibration of PVAL (Figure 4.34b). Thus, it can be concluded that some amount of PVAL still left in the PLA fibrils obtained from PLA/PVAL blend filaments, which were immersed in distilled water at room temperature up to ten days. However, PVAL seems to be almost completely removed from PLA/PVAL blend filaments at 60 °C, even only after 24 hours (Figure 4.34a). Moreover, there is no difference of the ATR-FTIR spectra measured from PLA fibrils, which were immersed in water at 60 °C in 1 day, 5 days, or 10 days. In conclusion, due to the very strong hydrogen bond between PLA and PVAL, it is impossible to get 100 % pure PLA fibrils without any presence of very small PVAL amount, even the PLA/PVAL blend filaments were immersed in water at 60 °C up to ten days. The ATR-FTIR in Figure 4.35b confirmed that PLA fibrils seem still to have very small amount of PVAL due to very slight bands in stretching region of OH and CH_2 groups in comparison to the neat PLA.

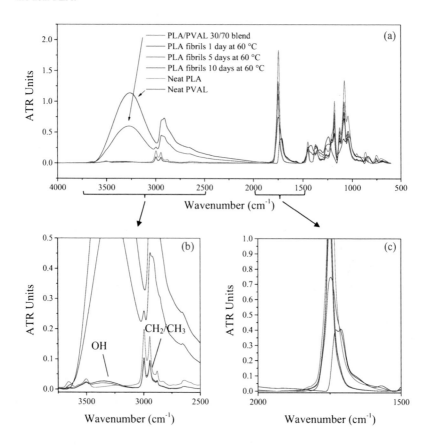

Figure 4.35 ATR-FTIR spectra of neat PLA, PVAL, PLA/PVAL 30/70 blend, and the PLA scaffolds obtained from PLA/PVAL blend filaments, which were immersed in water bath at 60 °C for 1, 5, and 10 days.

4.2.5 Mechanical properties of nanofibrillar PLA structures (PLA scaffolds)

Figure 4.36 demonstrates stress-strain curves of ten different scaffolds, which are prepared from drawn PLA/PVAL blend filaments with draw ratio DR=1.50 (Figure 3.16 and 3.17, section 3.3.10). Figure 4.37 shows the state of these scaffolds before and after tensile testing. It is seen that the scaffolds have a typical tensile strength of polymeric materials, but they have very little elongation at break (ca. 2 %). Furthermore, It is worth noting that the scaffolds, which are prepared from PLA/PVAL blend filaments without further off-line drawing (DR=1.0), are very brittle. They have almost no strength and elongation that could not be performed by a tensile testing device. These results confirmed that the off-line drawing is necessary to improve the mechanical properties of scaffolds. During drawing the PLA nanofibrils in PLA/PVAL blend filaments are more elongated, thinned, and oriented as presented in Figure 4.29 and 4.30 (section 4.2.3).

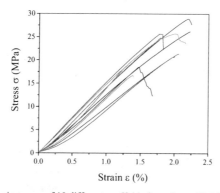

Figure 4.36 Stress-strain curves of 10 different scaffolds from drawn PLA/PVAL filaments

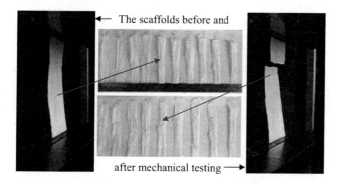

Figure 4.37 Scaffolds before and after mechanical testing

4.2.6 Discussion

Using the conventional melt spinning method of thermoplastic polymeric fibers, PLA/PVAL blend filaments (including nanofibrillar structures) were fabricated. The formation process of the in situ nanofibrillar morphology probably involves dispersion of the minor phases, single particle deformation, and coalescence process of elongated particles. The deformation and coalescence at the die entrance and inside the die are extremely important steps under melt extrusion [29]. In the melt spinning process, it is well known that the polymer mainly undergoes shear flow during compounding in twin-screw extruder and also during passing the capillary in the spinneret channel. After that, the elongational deformation along the spinline becomes more important. However, few studies have tried to study the morphological variations of polymeric blend systems after extrusion from the spinneret orifices into fibers [26, 55, 56]. These investigations focused on the morphology of the dispersed phase in blend samples at only two positions along the spinline: Blend samples at the die exit and the as-spun fibers taken from the bobbin. These studies showed that no fibrillar morphology or thick discontinuous fibrils of the dispersed phase was found in the blend extrudates (undrawn filaments) at the die exit, while thin endless fibrillar structures were observed in the as-spun blend fibers (blend filaments). It can be seen from Figure 4.12 that there was a significant difference of morphology of the dispersed PLA phase in PLA/PVAL blend extrudates and in PLA/PVAL blend filaments. While the PLA structures in PLA/PVAL blend extrudates showed the shape of spherical, ellipsoidal droplets or short fibrillar structures with typical diameters in microscale (~1 — 5 μm), the PLA structures in PLA/PVAL blend filaments appeared as uniform continuous long thin fibrils, their typical diameters here were in nanoscale (~30 — 200 nm). It can be assumed that the elongational deformation within the fiber formation zone strongly affected the morphology development of the dispersed phase in polymer blends along the spinline.

In the next two sections, the morphological evolution of the PLA/PVAL blend filaments along the spinline is investigated. Emphasis of the study is given on considering the filament parameters within fiber formation zone of the melt spinning process such as filament temperature, axial velocity gradient (axial strain rate) and tensile stress that influence the final state of deformation of the dispersed PLA phase. It is worth mentioning here that all melt spinning experiments in the next two sections 4.3 and 4.4 were carried out on the piston type melt spinning device, which was described in Figure 3.3 (chapter 3, section 3.2.2)

4.3 Characterization of PLA/PVAL monofilament profiles

Table 4.4 lists the different spinning conditions for the low-speed melt spinning process of PLA/PVAL 30/70 blend filaments. For instance, the mass flow rate is varied from 0.5 to 2.0 g·min^{-1} for the constant take-up velocity of 50 m·min^{-1} and the take-up velocity is altered from v=10 m·min^{-1} to 70 m·min^{-1} for the constant mass flow rate of Q=1.0 g·min^{-1}, As mentioned above, all these spinning processes were done on the self-constructed piston spinning device at IPF Dresden e. V.

Table 4.4 Spinning conditions with an extrusion temperature of 195 °C, D_0=0.6 mm, L/D=2

Conditions	Take-up velocity (m·min^{-1})	Volumetric flow rate (cm^3·min^{-1})	Mass flow rate (g·min^{-1})
A	50	0.393	0.5
		0.785	1.0
		1.178	1.5
		1.570	2.0
B	10		
	30	0.785	1.0
	50		
	70		

The following sections present the investigation of the filament profiles such as filament temperature, velocity, velocity gradient, diameter, tensile force, tensile stress, and apparent elongational viscosity using both experimental measurements and theoretical calculations.

4.3.1 Filament temperature profiles

Figure 4.38 shows the temperature profiles $T(x)$ of PLA/PVAL blend filament along the spinline for the take-up velocity of 50 m·min^{-1} with the mass flow rate of 1.0 g·min^{-1} over the distances from 2.5 cm to 150 cm. The grey cross symbols (×) represent the temperature values obtained using an infrared camera (chapter 3, section 3.3.6), the red points (•) (including deviation) are the emissivity corrected temperatures, which depend on the filament diameter (Figure 3.11, section 3.3.6), and the red dash curve (– – –) is the fitted temperature profile using an exponential decay function.

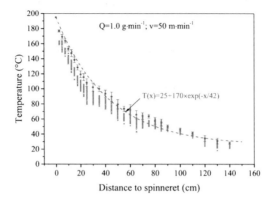

Figure 4.38 Filament temperature vs. distance for the take-up velocity of 50 m·min^{-1} and the mass flow rate of 1 g·min^{-1}: (×) uncorrected temperatures obtained using the infrared camera, (•) corrected temperatures, and (– – –) fitted temperature profile using an exponential decay function.

The main aim of this section is to determine the filament temperature variations along the spinline using an infrared camera, which were further used for understanding the effect of filament temperatures on the morphology development of PLA/PVAL blend filaments. Furthermore, only the fitted temperature profiles were presented to have better overview of the filament temperature profiles of seven different spinning conditions, which are listed in the above Table 4.4. More details of the experimental results, the fitted and simulated filament temperature profiles along the spinline are fully described and discussed in Appendix A.

Figure 4.39 shows the fitted filament temperature along the spinline versus the distance from the die exit for the constant take-up velocity of 50 m·min^{-1} with the different mass flow rates of 0.5, 1.0, 1.5, and 2.0 g·min^{-1}. It can be seen that the filament temperature profiles substantially depend on the mass flow rates. The filament cooled more slowly along the spinline when the mass flow rate increased at the constant take-up velocity.

In contrast, the filament temperature profiles seem to be independent of take-up velocities at the constant mass flow rate of 1.0 g·min^{-1} and have the similar temperature profiles corresponding to the same following fitting Equation 4.1 (Figure 4.40):

$$T(x) = 25 + 170 \times \exp\left(\frac{-x}{42}\right) \qquad (4.1)$$

where, $T(x)$ is the filament temperature in °C at any distance x in cm from the die exit.

It is worth mentioning here that the glass transition temperature T_g of PLA/PVAL filaments is assumed equal to T_g of PLA/PVAL blend obtained using DSC measurement. From Table 4.1 (section 4.1.1) it can be seen that the $T_{g,PLA}$=55 °C is higher than the $T_{g,PVAL}$=41 °C in PLA/PVAL 30/70 blend. Thus, the PLA solidifies first and the PLA/PVAL blend is totally solidified at the glass transition temperature of PVAL. Further analysis of the correlation between the filament temperature and morphology development along the spinline will be discussed in the section 4.4.

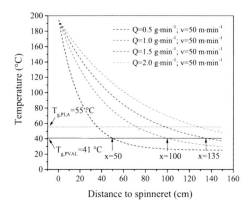

Figure 4.39 Filament temperature vs. distance for the take-up velocity of 50 m·min^{-1} and the different mass flow rates: 0.5, 1.0, 1.5, and 2.0 g·min^{-1}

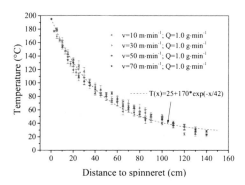

Figure 4.40 Filament temperature vs. distance for the different take-up velocities at constant mass flow rate of 1.0 g·min^{-1}

Results and discussion

4.3.2 Velocity and velocity gradient along the spinline

Figure 4.41 shows the velocity $v(x)$ profile of PLA/PVAL blend monofilament along the spinline for the take-up velocity of 50 m·min^{-1} with the mass flow rate of 1.0 g·min^{-1}. It should be noted that the grey cross symbols (×) represent for the velocity values obtained using a laser Doppler anemometry device (chapter 3, section 3.3.5) and the spinline-connected curve is the fitted curve by manual adjustment. In this section, it does not tend to present all experimental results of online velocity measurement, which are completely presented in the Appendix B. It is worth noting here that only the fitted values of filament velocity were shown to get better overview of filament velocity along spinline.

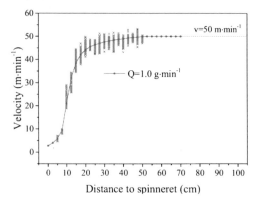

Figure 4.41 Velocity vs. distance for the specific melt spinning condition: v=50 m·min^{-1}; Q=1 g·min^{-1}

Figure 4.42a shows the manual fitted curves of the filament velocities versus distance to spinneret for the melt spinning processes with a constant take-up velocity of 50 m·min^{-1} and mass flow rates over the range from 0.5 g·min^{-1} to 2.0 g·min^{-1}. Figure 4.42b represents the axial velocity gradient of the filament velocities versus distance to spinneret. The axial velocity gradient or the axial strain rate (ASR) is used to characterize the deformation behavior of the material. The ASR is determined by taking the derivative of $v(x)$ with respect to x: $v'(x) = dv(x)/dx$. Thus, the ASR is not constant along the spinline due to the increase of velocity. It has a maximum value at position in which the velocity changes largest amount per time. From the Figure 4.42b, it is seen that the maximum value of the ASR decreases with increasing flow rate at the constant take-up velocity. At a higher flow rate, the ASR maximum region has a broader distribution and the ASR maximum peak slightly shifts towards longer distances from the spinneret.

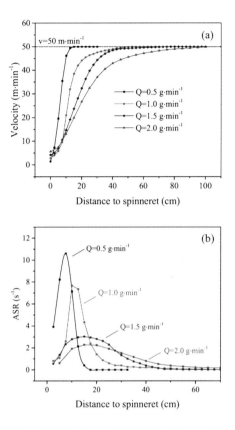

Figure 4.42 Velocity (a) and axial strain rate (ASR) (b) vs. distance for the melt spinning conditions: v=50 m·min^{-1}; Q=0.5, 1.0, 1.5, and 2.0 g·min^{-1}

Figure 4.43 shows the filament velocity and the ASR profiles along the spinline for the melt spinning processes with take-up velocities over the range from 10 m·min^{-1} to 70 m·min^{-1} at a constant mass flow rate of 1.0 g·min^{-1}. It can be seen from Figure 4.43b that the maximum value of the ASR increases with increasing take-up velocity. The region of the ASR maximum value did not change much and is located at a distance of ca. 10 – 13 cm from the spinneret for various take-up velocities with the exception of the take-up velocity of 10 m·min^{-1}. The ASR profile of the take-up velocity of 10 m·min^{-1} shows a maximum value at a region of ca. 6 – 9 cm.

Results and discussion

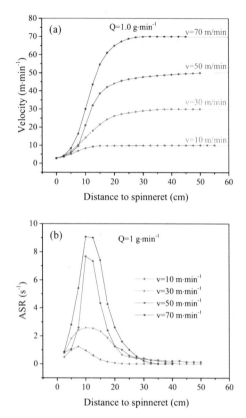

Figure 4.43 Velocity (a) and ASR (b) vs. distance for the melt spinning conditions: Q=1.0 g·min⁻¹; v= 10, 30, 50, and 70 m·min⁻¹

4.3.3 Filament diameter profiles

The filament diameter along the spinline was determined in two ways. First, the filament diameter $D(x)$ versus distance x from the die exit was obtained by measuring the diameter of the captured pieces of PLA/PVAL blends filament at different locations using a microscope. Second, the filament diameter $D(x)$ along the spinline was calculated from velocity data $v(x)$ and volumetric flow rate $\dot{V}$ by the Equation 4.2 below:

$$D(x) = \frac{2}{\sqrt{\pi}} \times \sqrt{\frac{\dot{V}}{v(x)}} \tag{4.2}$$

where, the filament velocity $v(x)$ at any distance x along the spinline was measured using the laser Doppler anemometry as analyzed in the above section 4.3.2.

Figure 4.44 demonstrates the distributions of the two courses of filament diameter along the spinline for the take-up velocity of 50 m·min^{-1} at the mass flow rate of 1.0 g·min^{-1}. The filament diameter decreases along the spinline and reaches a constant value of 141 µm at a region 25-50 cm from the die exit. In the region from the position near the die exit up to ca. 10 cm, the filament diameter rapidly drops with increasing filament velocity (Figure 4.41). It can be observed that the optically measured diameter profile well coincides with the calculated diameter profile from velocity data. This result allows one to use the calculated filament diameters for the all spinning conditions listed in table 4.4, because the calculation of the filament diameter using the Equation 4.2 is much faster and more convenient than that using the optical technique. This result also confirms that the fiber-capturing device did not affect the diameter of PLA/PVAL blend filaments. There is no additional stretching of the filaments and no changing of their shape. Therefore, it can be assumed that the original morphology of blend systems developed inside the filament was almost completely retained.

Figure 4.45 and Figure 4.46 show the filament diameter versus distance to spinneret for different mass flow rates at the constant take-up velocity of 50 m·min^{-1} and for the constant mass flow rate of 1.0 g·min^{-1} at different take-up velocities, respectively. It can be clearly seen from Figure 4.45 that the change in mass flow rate substantially effects on the termination of filament attenuation process. The filament diameters of lower mass flow rates attenuate much more rapidly than that of higher mass flow rates and they reach their final diameters at a location closer to the die exit. This is because that as the mass flow rate decreases, the filament reaches the solidification temperature faster (Figure 4.39).

In contrast, the take-up velocities almost did not have an affect on the termination of filament attenuation process. The filament diameters reach their final value at the similar distance from the die exit, because the change of filament temperature along the spinline has the similar behavior (Figure 4.40).

Results and discussion

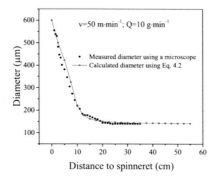

Figure 4.44 Filament diameter vs. distance for the spinning condition: Q=1.0 g·min^{-1}; v= 50m·min^{-1}

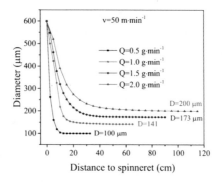

Figure 4.45 Filament diameter vs. distance for the spinning conditions: v= 50m·min^{-1}; Q= 0.5, 1.0, 1.5, and 2.0 g·min^{-1}

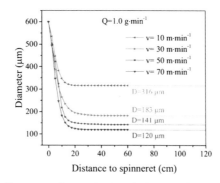

Figure 4.46 Filament diameter vs. distance for the spinning conditions: Q= 1.0 g·min^{-1}; v=10, 30, 50, and 70 m·min^{-1}

4.3.4 Tensile force, tensile stress, and apparent elongational viscosity profiles

In the melt spinning processes with low spinning speed up to ca. 750 m·min⁻¹, the surface tension force $F_{surf}(x)$, the air drag force $F_{aero}(x)$ and the inertial force are usually negligible as small as compared to the take-up force F_L and the gravitational force $F_{grav}(x)$ [153, 154, 225, 226]. In this study, therefore, **the tensile forces $F(x)$** are calculated using the force balance Equation 2.21 (chapter 2, section 2.3.1) that can be simplified as follows:

$$F(x) = F_{rheo}(x) \approx F_L + F_{grav}(x) \tag{4.3}$$

where F_L, is the take-up force. The gravitational force $F_{grav}(x)$, the weight of the filament at any distance x, can be calculated from the equation:

$$F_{grav}(x) = \int_x^L \varrho_P . g . \dot{V}/v(\tilde{x}) . d\tilde{x} \tag{4.4}$$

where g is the gravitational acceleration on earth ($g \approx 9{,}81\, m.s^{-2}$) and L is the take-up distance.

The tensile stress $\sigma(x)$ is given by the equation

$$\sigma(x) = F(x)/A(x) \tag{4.5}$$

where, $F(x)$ is the tensile force and $A(x) = \dot{V}/v(x)$ denotes the cross-sectional area of the filament.

Apparent elongational viscosity (APEV) $\eta_{app}(x)$ along the spinline was calculated by the following equation.

$$\eta_{app}(x) = \sigma(x)/v'(x) \tag{4.6}$$

where, $\sigma(x)$ is the tensile stress and $v'(x)$ is the derivative of $v(x)$ with respect to x. More details of theoretical analysis of elongational viscosity can be found in Ref. [227].

In this study, the take-up force $F_L = 1.9$ cN was determined at a distance of 180 cm below the die exit and near take-up device using a Tensiometer (section 3.3.7) for all spinning conditions are listed in the above Table 4.4. Within the measuring range of the Tensiometer, there is no different take-up force between spinning conditions that can be detected, because both the air drag force F_{aero} (estimated using Equations 2.25, 2.27 and 2.28) and inertial force

Results and discussion

F_{inert} (calculated using Equation 2.26) are too small in comparison to the total force and they do not have any effect on the total force. It is seen from Figure 4.47b that total forces of the air drag plus inertial force at a distance 180 cm below the die exit for different spinning conditions have minimum and maximum values of 0.00071 cN and 0.0071 cN, respectively. This result shows a considerable difference in force value between different spinning conditions, but this force value is too small in comparison to the take-up force $F_L = 1.9$ cN, determined using a Tensiometer. Therefore, it is simplify assumed that the take-up force F_L is constant for all spinning conditions. For the calculations of tensile force $F(x)$ (after Equations 4.3 and 4.4), tensile stress $\sigma(x)$ (after Equation 4.5), and apparent elongational viscosity $\eta_{app}(x)$ (after Equation 4.6), the measured and interpolated values of $v(x)$ were always used.

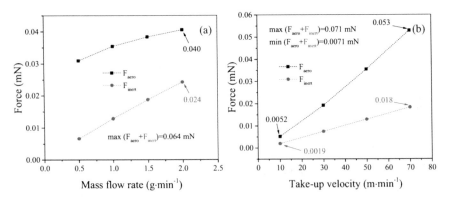

Figure 4.47 Air drag and inertial force at a distance 180 cm from the spinneret for different spinning conditions: (a) different mass flow rates and constant take-up velocity, (b) different take-up velocities and constant mass flow rate

Figure 4.48 and 4.49 show the distributions of the tensile force along the spinline for different spinning conditions. It can be seen that the tensile force is nearly constant along the spinline (Figure 4.48a and 4.49a). The tensile force slightly increases with increasing the mass flow rate at the constant take-up velocity of 50 m·min^{-1} (Figure 4.48b) or with decreasing the take-up velocity at the constant mass flow rate of 1.0 g·min^{-1} (Figure 4.49b). This tendency is due to the differences of filament diameter (Figure 4.45 and 4.46), which affect the gravitational force.

For each spinning condition, the filament diameter decreased along the spinline in the fiber formation zone (Figure 4.45 and 4.46). This decrease leads to the considerable increase of the tensile stress (Figure 4.50), although the variation of tensile force along the spinline is very small (Figure 4.48a and 4.49a). Table 4.5 lists the tensile stress at maximum axial strain rate (ASR) for different spinning conditions and Figure 4.51 plots the tensile stress at maximum ASR as a function of mass flow rate and take-up velocity. It can be found that the tensile stress at maximum ASR decreases rationally with the increase of the mass flow rate at the constant take-up velocity (Figure 4.51a) and there is a linear relationship between the tensile stress at maximum ASR and take-up velocity for the constant mass flow rate (Figure 4.51b).

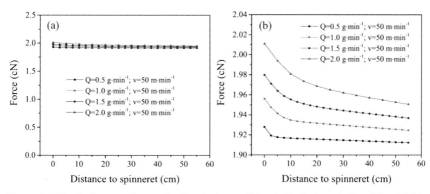

Figure 4.48 Tensile force vs. distance for the spinning condition A: Vertical y-axis from 0.0 to 2.5 (a), and 1.90 to 2.04 (b)

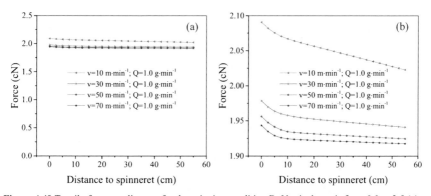

Figure 4.49 Tensile force vs. distance for the spinning condition B: Vertical y-axis from 0.0 to 2.5 (a), and 1.90 to 2.10 (b)

Results and discussion

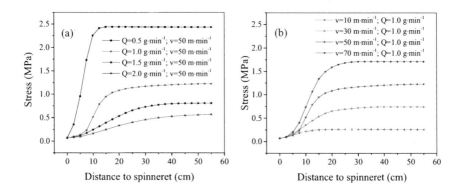

Figure 4.50 Tensile stress vs. distance for the spinning condition A (a) and B (b)

Table 4.5 Tensile stress at maximum axial strain rate (ASR) for different spinning conditions

Conditions	Take-up velocity (m·min^{-1})	Mass flow rate (g·min^{-1})	Tensile stress at maximum ASR (MPa)
A		0.5	1.73
	50	1.0	0.52
		1.5	0.39
		2.0	0.25
B	10		0.18
	30	1.0	0.35
	50		0.52
	70		0.74

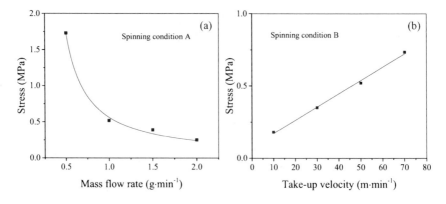

Figure 4.51 Tensile stress at maximum ASR for different spinning conditions A (a) and B (b)

Figure 4.52 shows the calculated apparent elongational viscosity (APEV) of PLA/PVAL blend filaments versus distance from spinneret for different spinning conditions. It is seen that the APEV keeps an almost constant value at the first 10 centimeters. It then rapidly increases and goes to infinity, where the attenuation process is finished (Figure 4.52a and 4.52b). It is seen from Figure 4.52a that as the mass flow rate increases the APEV reaches an infinite value more slowly, because the filament cools more slowly along the spinline (Figure 4.39). Figure 4.52b shows an interesting relationship between the APEV and the take-up velocity. In general, as the take-up velocities decreases, the APEV becomes infinite at the shorter distances from the spinneret. However, this tendency is not found at a take-up velocity of 70 m·min^{-1}. The APEV at this take up velocity goes to infinity even earlier than that of take-up velocities of 30 and 50 m·min^{-1}. An explanation of this behavior is that the melt spinning process with take-up velocity of 70 m·min^{-1} is faster terminated than that of with take-up velocity of 30 and 50 m·min^{-1}. This agrees well with the velocity results that have already shown in Figure 4.43a.

As mentioned above, the APEV depends on the ASR after Equation 4.6. It depends also on filament temperature: as the filament temperature decreases, the APEV increases. The temperature profiles in Figure 4.40 show that there are no differences of the filament temperature between various take-up velocities, because they were fitted from the same Equation 4.1. However, the experimental results of filament temperature in Appendix A show a little difference. The filament in the melt spinning process with the take-up velocity of 70 m·min^{-1} is quenched a little earlier than that of with take-up velocities of 30 and 50 m·min^{-1}.

It is worth here pointing out from Figure 4.52c and 4.52d that the APEV has a minimum value at several centimeters below the spinneret, where the maximum ASR values are found (Figure 4.42b and 4.43b). This phenomenon occurs because the APEV is inversely proportional to the ASR after Equation 4.6. The similar tendency was also found in the Ref. [228].

4.3.5 Discussion

In the melt spinning process, the fiber formation zone is defined as the distance L_S from the die exit x_0 ($x = 0$) to the solidification point x_s ($x = s$), where the fiber becomes solid. In this zone, the morphological development of the blend filament occurs and it is affected by the courses of spinning parameters as analyzed above. Below solidification point no further deformation of the filament occurs, i.e. the process of morphological development of blend filaments no more happens.

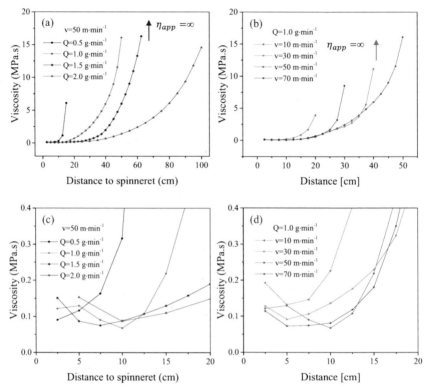

Figure 4.52 Apparent elongational viscosity vs. distance for the spinning condition A (a and c) and spinning condition B (b and d)

In principle, at the same time t_s, the filament velocity should reach the take-up velocity and filament temperature should cool down to its glass transition temperature T_g at the same solidification distance L_S. Figure 4.53 plots the solidification distance versus mass flow rate and take-up velocity from the fitted velocity (section 4.3.2) and fitted temperature profiles (section 4.3.1). For all analyses below, it is assumed that T_g of the blend filaments from fitted temperature profiles equal to $T_{g,DSC}$ obtained using DSC method[1]. It can be seen that the distance L_{Sv} in which the filament velocity reaches its take-up velocity, i.e. the solidification

---

[1] There is a difference of glass transition temperature between T_g and $T_{g,DSC}$, because thermal behavior of polymers in DSC measurement and in the melt spinning process appears in two different equilibrium and non-equilibrium state, respectively [175].

Results and discussion

distance obtained from velocity profiles, is always shorter than the solidification distance L_{ST} obtained from the fitted temperature profiles ($L_{Sv} < L_{ST}$). This interesting phenomenon may occur for the low melt spinning speed up to 70 m·min^{-1}. It may be that filament needs more time to go to its glass transition temperature than to reaches its take-up velocity or very near take-up velocity.

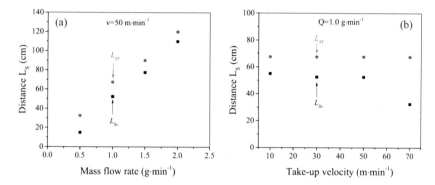

Figure 4.53 Solidification distance from fitted temperature profiles L_{ST} (the red-coloured filled circle (●)) and fitted velocity profiles L_{Sv} (the black-coloured filled square (■)) vs. mass flow rate (a) and vs. take-up velocity (b)

The following Equation 4.7 and 4.8 was used to calculate the time t_{Sv} required for the filament velocity reaches its take-up velocity and the time t_{ST} needed for the filament temperature goes to its glass transition temperature, respectively.

$$t_{Sv} = \int_{x_0}^{x_{Sv}} \frac{1}{v(x)} dx \qquad (4.7)$$

$$t_{ST} = \int_{x_0}^{x_{ST}} \frac{1}{v(x)} dx \qquad (4.8)$$

where, x_{Sv} and x_{ST} are the solidification points from the fitted velocity profiles and the fitted temperature profiles.

The time difference at $\Delta t = t_{ST} - t_{Sv}$ changes a little over a range of 0.12 to 0.21 second for the constant take-up velocity and different mass flow rates (Figure 4.54a). It can also be seen from Figure 4.54b that there is no substantial time difference Δt for the take-up velocity 30, 50, and 70 m·min^{-1}, but except for the very slow take-up velocity of 10 m·min^{-1}. The final filament fineness of the take-up velocity of 10 m·min^{-1} is 318 μm almost twice as large as that of the take-up velocity of 30 m·min^{-1}. This means that the filament at the take-up velocity 10 m·min^{-1} need much more time that at the higher take-up velocities for cooling down to its glass transition temperature although the filament velocity has already reached its take-up velocity.

The difference between the solidification distance L_{Sv} and L_{ST} may be caused by some following reasons: (1) It may be due to the axial heat conduction effect of the filament along the spinline and non-uniform radial temperature through the cross-section of the thick filament, i.e. the outer layer of the filament cooler than the inner layer. (2) There may occur the telescoping effect or "necking" during stretching the filament within fiber formation zone. (3) And it is due to the fact that the two measurement methods (laser Doppler velocimetry and infrared thermography) are differences themselves.

In summary, by considering the length of deformation zone, it is reasonable to assume that the fiber formation zone is from the die exit to the point at the filament velocity reaches its take-up velocity. The end of deformation point equals the first point along the spinline where filament velocity reaches its take-up velocity or very near take-up velocity. In this zone, the morphological development of PLA/PVAL is mainly deformed by the increase of velocity and tensile stress.

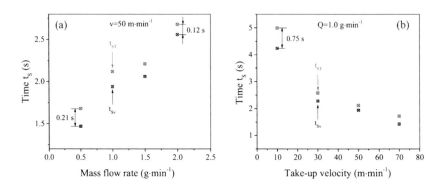

Figure 4.54 Solidification time vs. mass flow rate (a) and take-up velocity (b)

Results and discussion

4.4 Morphology development of PLA/PVAL blends in an elongational flow

4.4.1 Morphology development of PLA/PVAL 30/70 blends along the spinline[1]

This section presents visually the morphological development of PLA/PVAL 30/70 blend under the specific spinning condition with the take-up velocity of 50 m·min^{-1} and the volumetric flow rate of 0.785 cm^3·min^{-1} (mass flow rate of ca. 1.0 g·min^{-1}).

Figure 4.55 gives a schematic view of the positions (P0 – P8) where the blend samples were taken to investigate their morphology (blend granules before extrusion and melt spinning, position 0 (P0); captured blend extrudates along the spinline, positions 1 – 7 (P1 – P7); blend filaments, position 8 (P8)).

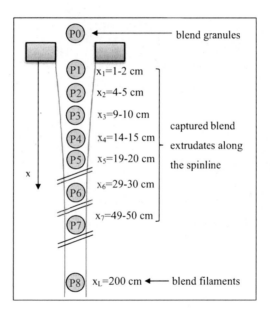

Figure 4.55 Schematic view of a monofilament and locations of the captured samples

Figure 4.56 – 4.60 present the morphology of the dispersed PLA phase after removing PVAL matrix from PLA/PVAL blends at different locations, obtained by SEM. Figure 4.56 and 4.57

---

[1] Parts of this section have been published in: N. H. A. Tran, H. Brünig, R. Boldt, and G. Heinrich. *Polymer* 55(24), 6354-6363 **(2014)**

show the morphology of dispersed PLA phase from the PLA/PVAL blend pellets, obtained from internal mixer and twin-screw extruder (section 3.2.1), respectively, before extrusion and melt spinning (Position 0, P0). It is seen that the dispersed PLA droplets in blend granules prepared using the twin-screw extruder are finer and more elongated than those of prepared using the internal mixer. It is well known that the mixing of polymer blends in the twin-screw extruder is more effective than that in the internal mixer [229, 230]. Therefore, the blend granules obtained from the twin-screw extruder were used as the initial granules for the present study.

Figure 4.56 SEM images of the dispersed PLA phase from PLA/PVAL blend granules (P0), prepared using internal mixer after removing the PVAL matrix: (a) scale bar 10 µm; (b) scale bar: 1 µm

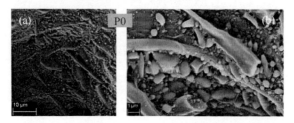

Figure 4.57 SEM images of the dispersed PLA phase from PLA/PVAL blend granules (P0), prepared using twin-screw extruder after removing the PVAL matrix: (a) scale bar 10 µm; (b) scale bar: 1 µm

Figure 4.58 gives the SEM images of the dispersed PLA phase from PLA/PVAL extrudates for the position near the die exit (P1, 1 — 2 cm below the die exit). Under the effect of shear flow during passing the capillary of the spinneret die (Figure 3.3), the spherical/ellipsoidal PLA domains (Figure 4.56 and 4.57) were deformed and ruptured into smaller particles in randomly distribution (Figure 4.58). The formation of such structures agrees well with the breakup mechanisms of dispersed phases in blends, which have been described and summarized by Sundararaj [138]. Such breakup mechanisms can happen for the PLA/PVAL blend because of the high difference of the viscosity levels between the two polymers as shown in Figure 4.18. The viscosity ratio $\eta_r = \eta_{PLA}/\eta_{PVAL}$ is always less than 1 over the shear rate range from 0.1 to 100 rad·s^{-1} (Figure 4.18b).

Figure 4.58 SEM images of the dispersed PLA phase from PLA/PVAL extrudates at the location near the die exit (P1) after removing the PVAL matrix: (a) scale bar: 10μm; (b) scale bar: 1μm.

Figure 4.59 shows the PLA nanofibrils from PLA/PVAL blend filaments taken from the bobbin (P8) after removing PVAL matrix. Comparing Figures 4.58 and 4.59, it can be seen that there is a huge difference of PLA structures between the position 1 (P1) and position 8 (P8). During melt spinning, under the take-up tension within the fiber formation zone, the molten polymeric extrudate has to undergo uniaxial elongational flow for stretching purpose. The dispersed PLA phase is formed into long continuous nanofibrillar structures in PLA/PVAL blend filaments (Figure 4.59). A number of studies [128-130, 143, 172, 175, 231, 232] have quite clearly shown that spinning orientation develops gradually along the spinline. Elongational flow significantly affects the morphology and seems to be more efficient in deforming the dispersed phase than shear flow. The formation of fibrillar structures is governed by the local velocity or the local tensile stress within the fiber formation zone [129].

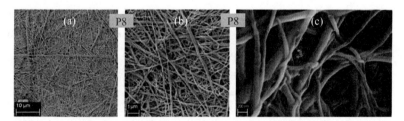

Figure 4.59 SEM images of PLA nanofibers from PLA/PVAL blend filaments (P8) after removing the PVAL matrix, spinning speed: 50 m·min^{-1}: (a) scale bar: 10μm; (b) scale bar: 1μm; (c) scale bar: 200 nm

Figure 4.60 presents the morphological development of PLA/PVAL blend extrudates at different locations along the spinline. The left and the middle columns display the SEM images[1] describing the morphological development of the dispersed PLA phase after dissolving the PVAL matrix. The right column shows the cross-sectional surfaces of PLA/PVAL blend extrudates after etching the dispersed PLA phase. It is noting that scale bar for the left column is 10 µm, scale bars for the middle and right columns are 1 µm.

It is seen that the size and the shape of the dispersed PLA phase rapidly changes in the region from the position near the die exit on the first centimeters (P1) (Figure 4.58) up to a distance of ca. 15 (P4) to 20 cm (P5) from the die exit (Figure 4.60g — h and 4.60j — k). In this region, the filament velocity increased rapidly along the spinline (Figure 4.41, section 4.3.2). The maximum velocity gradient was located at a distance of ca. 8 — 10 cm (Figure 4.42b — 4.43b, the red curve) where the morphology of dispersed PLA phase changed from short fibrils or rod-like structures (Figure 4.58) to continuous long thin fibrils (Figure 4.60g — h, P4). From these observations, there is enough evidence that the velocity gradient may be correlated to the deformation of the dispersed PLA phase in the PVAL matrix.

It is seen from the above section 4.3 that the filament temperature $T(x)$, the filament velocity $v(x)$, the filament diameter $D(x)$, the tensile stress $\sigma(x)$ and the apparent elongational viscosity $\eta_{app}(x)$ vary along the spinline. These parameters play an important role in the development of filament morphology and help to explain the fibrillation process of the dispersed phase in polymer blends. Based on all the SEM images (Figure 4.57 — 4.60) and the profiles of filament velocity, axial velocity gradient, filament diameter, tensile stress, temperature and apparent elongational viscosity along the spinline for the specific spinning condition with the take-up velocity of 50 m·min^{-1} and the mass flow rate of 1.0 g·min^{-1} (all the red colour curves in Figures 4.38 — 4.52, section 4.3), the mechanism of fibrillation process of dispersed PLA phase in PLA/PVAL blend extrudates can be presented schematically as shown in Figure 4.61. It can be separated in two-mechanisms: First, extrusion and shear flow in the capillary die (Figure 4.61, zone A), second, extension and elongational flow within the fiber formation zone (Figure 4.61, zone B, C, and D).

The first mechanism is dominated by the shear flow. In the capillary die (Figure 4.61, zone A), the spherical or ellipsoidal dispersed PLA phase (Figure 4.58 and Figure 4.61, P0) may erode at the surface, stretch and breakup in the flow direction, and spit out via a tip streaming mechanism [138]. Consequently, a random morphology of the dispersed PLA phase was observed at the die exit (Figure 4.58 and Figure 4.61, P1). The formation of such structures

---

[1] The original SEM images are completely presented in the Appendix C

was also earlier observed by Pesneau et al. [52], when they studied the morphology development of dispersed polyamide (PA) in polypropylene (PP) in a capillary die.

Although the thin-filament-model (Equations 4.1 — 4.6) is not able to evaluate the radial resolution of velocity, temperature, and stress fields. The following qualitative analysis, after [146, 175], can be appropriate to visualize a conceptual model of the second mechanism. This model can be divided into three steps or zones B, C and D (Figure 4.61).

On the first very short distances below the die exit (x ≈ 0 — 3 cm) (Figure 4.61, Zone B), the transition from shear to elongational flow and from a confined to a free-surface flow occurs. In this zone B, the cooling down of the filament starts and gives rise to the symmetric radial distribution of viscosity on the cross-section of the filament due to the temperature difference between inner and outer layers of filament volume element (Figure 4.62). Viscosity is a strong function of the temperature. It increases as the temperature decreases. The symmetrical radial distribution of viscosity leads to the possibility of simultaneously occurring of symmetrical radial velocity- and/or stress- distribution. However, it is seen that the axial filament velocity only slightly changes and the axial filament velocity gradient along the spinline in this zone is still at a low level (Figure 4.41, Figure 4.42b). Thus, the temperature difference between inner and outer layer of the filament volume element has very little effect on the radial velocity difference between layers on the cross-section of the filament. In this zone B, the blend extrudate is in molten state at a higher temperature than its melting point (Figure 4.38). Combining these analyses with the SEM images for the Position 1 (Figure 4.58 and Figure 4.61, zone B), it seems to be evident that the fibrillation process did not happen.

In the zone C (x ≈ 3 — 20 cm), where the axial filament velocity increases rapidly and reaches a value of 45 m·min^{-1}, that is 90 % of the take-up velocity, at a distance of ca. 20 cm from the die exit (Figure 4.41 and Figure 61, Zone C). Due to the increase of the velocity, the ASR significantly increases and reaches a maximum value of ca. 7.7 s^{-1} at 10 cm, and then decreases to 1.4 s^{-1} at 20 cm (Figure 4.42b and 4.43b, the red colour curves). The filament diameter rapidly drops from 500 μm at 2.5 cm to 150 μm (near the final filament diameter of 141 μm) at 20 cm from the die exit. The filament temperature exponentially decreases and has a temperature of 130 °C at 20 cm below the die exit (Figure 4.38). The filament temperature at a distance of 20 cm from the die exit is much higher than its glass transition temperature. Thus, the filament is still in viscoelastic state. As the temperature decreases, the apparent elongational viscosity of polymer blend slightly increases up to ca. 0.71 MPa.s (Figure 4.52, the red colour curves). Due to the rapid diameter attenuation of the filament along the spinline, there may be almost no temperature difference between layers on the filament cross-section. Thus, the radial viscosity of filament becomes more uniform. As a result, it is assumed that there is almost no difference of velocity or tensile stress between the layers through the cross-section of the filament anymore. Consequently, all layers through the filament cross-section are more elongated and oriented parallel to the spinline due to the continuous increasing of axial filament velocity (Figure 4.61, Zone C). In this zone, tensile stress also increases very rapidly (Figure 4.50, the red colour curves). This allows the filament to retain the shape while the filament is in the quenching stage. From these analyses of filament profiles and the SEM images for the Positions 2 to 4 (Figure 4.60a – i), it is clearly seen that the formation of fibrillar structures takes places mainly in the region from the first centimeters to a distance of ca. 20 cm from the die exit, where the polymeric blend filament is converted from the molten to viscoelastic state, the filament velocity and diameter reaches their near final values. The ASR of the filament is considered as the most essential parameter to deform and coalesce the dispersed PLA phase from rod-like to nanofibrillar structures in such high viscosity ratio systems like PLA/PVAL blends.

In the zone D (x ≈ 20 – 50 cm), the axial filament velocity increases very slowly and reaches its final take-up velocity of 50 m·min^{-1} at ca. 50 cm from the die exit. The ASR of filament slightly decreases, keeps almost constant at a very low value and becomes zero at 50 cm below the die exit, where the filament velocity reaches the constant take-up velocity. Due to the very slowly increases of the axial filament velocity (from 45 to 50 m·min^{-1} for the 30 cm distance), the filament diameter does not decrease much, only 9 micrometers (from 150 to 141 µm). In this zone D, the filament temperature further exponentially decreases and reaches the temperature of ca. 75 °C at a distance of 50 cm from the die exit. At this point, the filament temperature should cool also down to the glass transition temperatures of PVAL $T_{g,PVAL}$ = 41 °C and PLA $T_{g,PLA}$ = 55 °C in PLA/PVAL 30/70 blends, obtained using DSC measurement as shown in Table 4.1, section 4.1.1. However, the glass transition temperature of the blend extrudate obtained using DSC must be not the same value as that of blend extrudate during melt spinning. Because while the heat capacity of polymers obtained using DSC, is determined by the analyzing its equilibrium melting behavior, the thermal behavior of polymers in the melt spinning process appears in non-equilibrium state [175]. Furthermore, as discussed in the above section 4.3.5, even if the filament temperature does not reach its glass transition temperature, the filament deformation should almost finish its deformation process at the point, where the filament velocity reaches its take-up velocity or very near take-up velocity. In this zone D, the apparent elongational viscosity rapidly increases and goes to infinity at this distance of 50 cm from the die exit, because the ASR goes to zero (after Equation 4.6, section 4.3.4). From the SEM images for the Positions 5 to 8 (Figure 4.60j – r), it is concluded that nanofibrils, which were formed in the zone C, are further extended into finer fibrils under the effect of the continuous increasing of the tensile stress at a high value of ca 1.22 MPa (Figure 4.50a and 4.50b, the red colour curve).

Below the point at which the filament reaches its take-up velocity (x > 50 cm) or its glass transition temperature T_g, the filament temperature further decreases and finally becomes equal to the room temperature at distances between 140-150 cm from the die exit where the filament is fully solidified (Figure 4.38). There is no viscous flow deformation below T_g, while the filament is solidifying. The processes of deformation, coalescence, and break-up of fibrils may not occur anymore. Consequently, the nanofibrils with diameters ranging from 30 nm to 200 nm (an average diameter of 60 nm) were obtained (Figure 4.59).

In summary, the results show that during melt spinning, the rod-like structures of dispersed PLA phase in PLA/PVAL blend is stretched and coalesced to form continuous long thin nanofibrils in fiber formation zone. Although the present study has been limited for the special spinning condition with the take-up velocity of 50 m·min^{-1} and the mass flow rate of 1.0 g·min^{-1}, it confirms that melt spinning is an effective method to form dispersed phase into continuous fibrillar structures due to the presence of an elongational flow. The increase of the velocity and tensile stress in the fiber formation zone are mainly responsible for the stretching and coalescence of the dispersed PLA phase into nanofbrils in the PVAL matrix of polymer blends.

However, there are still many open questions to explore: Can the dispersed PLA phase be deformed from short rod-like or ellipsoidal structures in micro-scale into continous fibrillar structures in nano-scale along the spinline under different mass flow rate and/or take-up velocity? Are the sizes (lengths and diameters) of the final dispersed PLA nanofibrils caused only by the deformation of their initial sizes or by the combination of the deformation, coalescence, and break-up processes? Can the changes in spinning conditions control the deformation, coalescence, and break-up processes of the dispersed PLA phase in PLA/PVAL 30/70 blend extrudates within fiber formation zone? The following section 4.4.2 attempt to answer these questions.

Results and discussion

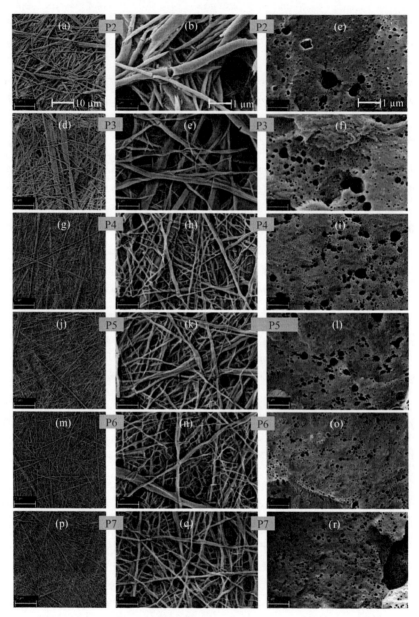

Figure 4.60 SEM images of PLA structures (left and middle columns) after removing PVAL matrix and PLA/PVAL cross-section after etching dispersed PLA phases (right column) at different locations along the spinline: P2 (a – c), P3 (d – f), P4 (g – i), P5 (J–L), P6 (m – o), and P7 (p – r).

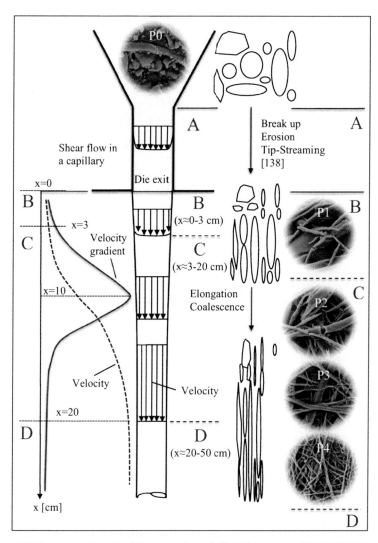

Figure 4.61 A conceptual model of the mechanisms of fibrillation process of PLA/PVAL blends in elongational flow along the spinnline under special spinning conditions D_0= 0.6 mm with L/D_0=2, Q=1.0 g·min^{-1} ($\dot{V}$=0.78 cm^3·min^{-1}), v=50 m·min^{-1}, T=195 °C.

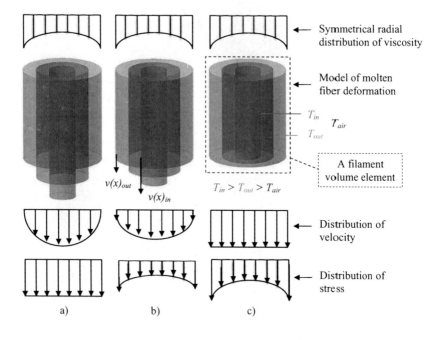

Figure 4.62 Schematic drawing three representative possible models of molten fiber deformation on the very first distances just below the die exit (x ≈ 0 − 3 cm) at symmetrical radial viscosity distribution: (a) symmetrical radial velocity distribution and uniform radial stress distribution, (b) symmetrical radial velocity and stress distribution, (c) uniform radial velocity distribution and symmetrical stress distribution, modified after [146, 175]

4.4.2 Controlling the micro- and nanofibrillar PLA morphology

This section presents in more details the morphology development of PLA/PVAL 30/70 blend extrudates/filaments at different locations along the spinline (Figure 4.55) for different spinning conditions (Table 4.4). The final nanofibrillar morphology of dispersed PLA phase in PLA/PVAL blend filaments can be controlled by changes in take-up velocities and mass flow rates.

Morphology of PLA/PVAL blend granules (P0, Figure 4.55)

Figure 4.63 shows the SEM images[1] of the PLA/PVAL 30/70 blend pellets obtained from the twin-screw extruder (section 3.2.1) before extrusion and melt spinning on the piston melt spinning device. Figure 4.64 presents frequency-distribution histograms of about 174000 dispersed PLA droplets in the fractured area of 114.7×74.1 µm^2 of a SEM image. For completely immiscible polymer blends systems, the size of the dispersed phase exhibits often log-normal behavior[2] [233-236]. However, for the partial miscible PLA/PVAL systems with the presence of the H-bonding as discussed above in the section 4.1, the fitting of the log-normal distribution seems to be not suitable, because the cumulative PLA droplet size distributions could not be fitted to a reference line as shown in Figure 4.65c. This observation agrees well with that reported by Harrats [233], who found the log-normal distribution does not apply for the immiscible polymer blends with changes in interfacial energy or compatibilization (copolymer formation).

It is seen from Figure 4.64 that the distribution of circular equivalent diameters (CEDs) of PLA droplets is very broad and the CEDs range from ca. 0.03 to 12.1 µm with an average number CED of 0.09 µm. The distribution of CED of PLA droplets can be roughly divided in two groups, in which CEDs are smaller and bigger than 0.5 µm. There are 98.6 % the number of droplets has diameter up to 0.5 µm (Figure 4.65a and 4.65c), only 1.4 % has diameter larger than 0.5 µm (including 1 % droplets has diameter within 0.5 to 1 µm, and a few droplets ca. 0.4 % has diameter larger than 1 µm) (Figure 4.65b and 4.65c). Furthermore, while the droplets having CEDs ≤ 0.5 µm are almost perfectly spherical, the droplets having CEDs > 0.5 µm are not a spherical but rather an ellipsoidal, cylindrical or elongated shape (Figure 4.63a–c). The calculated average circularity C_n of droplets with CEDs < 0.5 µm and CEDs > 0.5 µm is ca. 0.95 and 0.78, respectively (Figure 4.63d–f). An explanation might be that the small droplets are deformed more difficult than the large droplets under shear flow during mixing in twin-screw extruder. The large droplets tend to deform and then break up into smaller ones. It is worth pointing out here that although there are few large droplets, the volume of these droplets is considerable. The volume contributions may be more informative than the CED distribution. But, in the present study, the volume distributions could not be determined due to the ellipsoidal, cylindrical or elongated shape of the PLA droplets.

---

[1] Original SEM images are presented in the Appendix C.1
[2] The log-normal distribution is defined as the frequency of size versus the log of diameter of droplets results in a normal distribution

Results and discussion

In principle, the largest diameter of PLA droplet D may exist under simple shear flow in Newtonian and polymer blend systems can be estimated using Taylor's theory [131] (Equation 4.9) and Wu's correlation [237] (Equation 4.10), respectively:

$$D_{Taylor} = \frac{4\Gamma(\eta_r + 1)}{\dot{\gamma}\eta_m \left(\frac{19}{4}\eta_r + 4\right)} \quad \text{for } \eta_r < 2.5 \quad (4.9)$$

$$D_{Wu} = \frac{4\Gamma\eta_r^{-0.84}}{\dot{\gamma}\eta_m} \quad \text{for } \eta_r < 1.0 \quad (4.10)$$

where, $\Gamma = 1.6$ mN·m^{-1} is the interfacial tension[1], $\eta_r = \eta_d/\eta_m$ is the viscosity ratio between the dispersed phase viscosity η_d and the matrix viscosity η_m, and $\dot{\gamma}$ is the shear rate. η_r and η_m are obtained using oscillatory rheometer (Figure 4.18). It is worth noting here that the viscosities η_r and η_m depend on shear rate. Shear rate $\dot{\gamma}$ during mixing in twin-screw extruder at the screw speed of 100 rpm is discussed as follows.

In twin-screw extruder, polymers are subjected to complex shear and elongational deformations, and complex temperature profiles along the various zones of the extruder barrel. Therefore, it is difficult to characterize the type and magnitude of the strain rate in an extruder by a single number [237]. However, an "average shear rate" and "effective shear rate" may be used to approximately determine shear rate. Suparno [238] has theoretically predicted that the average shear rate linearly increases with the increase of screw speed and he found that the average shear rate is ca. 50 rad·s^{-1} at the screw speed of 100 rpm. Wu [237] and Burkhardt et al. [239] pointed out that the effective shear rate G in s^{-1} is almost equal to the crew speed n in rpm: $G \cong n$. In the present study, a crew speed of 100 rpm was used. Thus, it is assumed that during mixing of the PLA/PVAL blends with the screw speed of 100 rpm, the effective shear rate is 100 s^{-1} and the shear rate may vary from a few rad·s^{-1} to 100 rad·s^{-1} within mixing zone in twin-screw extruder.

Figure 4.66 gives the comparison of the measured maximum diameter $D_{Measurement}$ with the calculated maximum diameter of PLA droplets after Taylor's theory D_{Taylor} (Equation 4.9) and Wu's correlation D_{Wu} (Equation 4.10) within shear rate range from 0.01 to 100 rad·s^{-1}. It is seen that the $D_{Measurement}$ is always larger than the D_{Taylor}. It has almost the same value of ca. 12 μm with D_{Wu} at the shear rate of 1 rad·s^{-1} (Figure 4.66b). As discussed above, the effective shear rate for the screw speed of 100 rpm is definitely larger than 1 rad·s^{-1}. Thus, it

---

[1] The determination of interfacial tension between PLA and PVAL is presented in Appendix D.1

Results and discussion

can be clearly seen from Figure 4.66b that the $D_{Measurement}$ is also always larger than the D_{Wu}. It is worth pointing out here that the Taylor's theory was employed for Newtonian liquid mixtures independent on the blend ratio and Wu's correlation was used for two specific blend systems: ethylene-polypropylene rubbers as a dispersed phase in PA66 and PET matrix with the concentration of the dispersed phase of 15 %. In the present study, the used concentration of the dispersed phase is 30 % much higher than that used in Wu's correlation. It is well known that coalescence can occur at values of dispersed phase as low as 1 % and it increases with increasing dispersed phase content [126]. Due to the coalescence of the high PLA content (30 %) in PLA/PVAL blend, the measured maximum diameter is larger than the calculated maximum diameter after the Wu's correlation ($D_{Measurement} > D_{Wu}$).

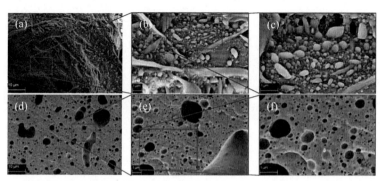

Figure 4.63 SEM images of the PLA/PVAL 30/70 blend pellets after removing PVAL matrix (a–c) and etching PLA dispersed phase (d–f) with scale bars for the left, middle, and right column are 10, 1, and 1 µm, respectively.

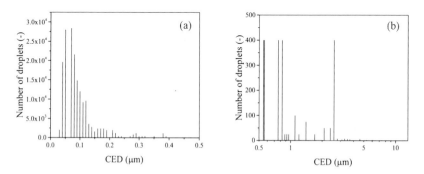

Figure 4.64 Frequency-distribution histograms vs. circular equivalent diameter (CED) over the range of CED up to 0.5 µm (a) and CED from 0.5 to 13 µm (b)

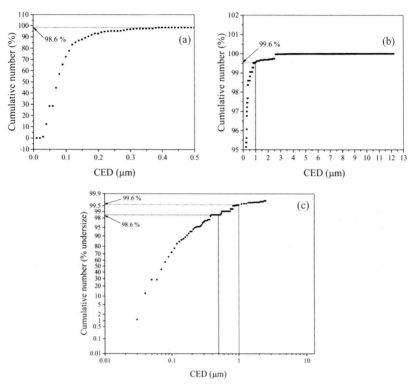

Figure 4.65 Cumulative number percentage vs. CED (a and b) and log-normal distributions (c)

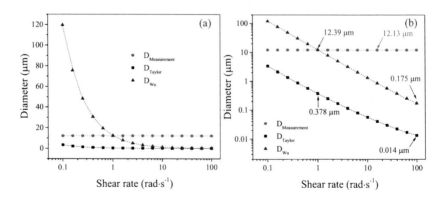

Figure 4.66 Comparison of the maximum diameter of PLA droplet vs. shear rate range: (a) linear scale and (b) logarithmic scale for the y-axis.

Results and discussion

Morphology of PLA/PVAL blend extrudates

Figure 4.67 presents the SEM images[1] of the cross-sectional surfaces of PLA/PVAL blend extrudates after etching the dispersed PLA phase. Figure 4.68 shows the SEM images of dispersed PLA phase after removing the PVAL matrix from the PLA/PVAL blend extrudates without stretching for various mass flow rates or extrusion rates. It is clearly seen that as the mass flow rate increases, the maximum circular equivalent diameter (CED) of PLA droplet increases (Figure 4.67 — 4.69). However, the number of PLA droplets (Figure 4.70a) and the number-averaged CED[2] $\bar{d}_{CED}$ decrease (Figure 4.69) with the increase of the mass flow rate. These results can be explained due to the coalescence process during extrusion through a convergent capillary die as follows.

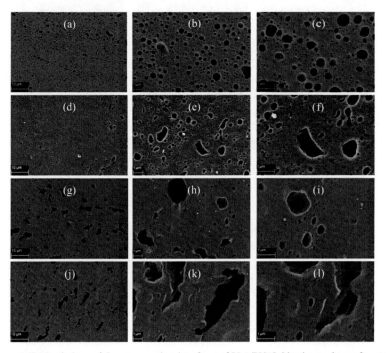

Figure 4.67 Morphology of the cross-sectional surfaces of PLA/PVAL blend extrudates after etching dispersed PLA phase for the various mass flow rates (extrusion rate) of 0.5 (a–c), 1.0 (d–f), 1.5 (g–i), and 2.0 g·min^{-1} (j–l): The scale bars from the left to right column are 10, 1, and 1 μm

---

[1] Original SEM images are presented in the Appendix C.2
[2] The calculation of number-averaged CED $\bar{d}_{CED}$ was presented in section 3.3.11

Results and discussion

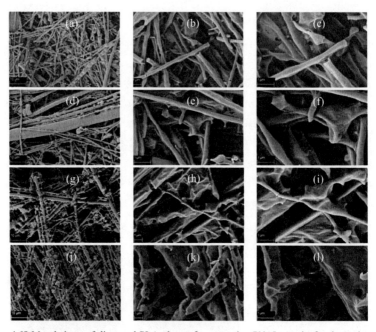

Figure 4.68 Morphology of dispersed PLA phase after removing PVAL matrix for the various mass flow rates (extrusion rate) of 0.5 (a–c), 1.0 (d–f), 1.5 (g–i), and 2.0 g·min^{-1} (j–l) with scale bars from the left to right column are 10 μm, 1 μm, and 1 μm, respectively.

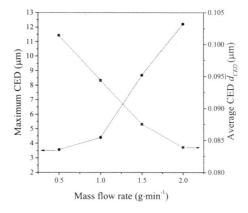

Figure 4.69 Maximum and number average CED $\bar{d}_{CED}$ vs. mass flow rate for the cross-section of PLA/PVAL blend extrudates after etching dispersed PLA phase (measured from Figure 4.67)

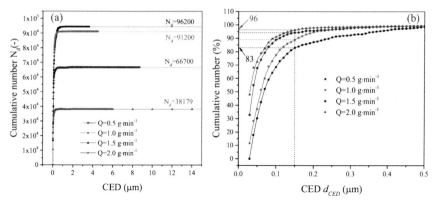

Figure 4.70 Cumulative number vs. CED (a), Cumulative number percentage vs. CED (b)

When the PLA/PVAL blends are forced to flow through a convergent capillary die in piston melt spinning device (Figure 4.71b), they are subjected to only normal stress (force acting perpendicular to the surface) in the zone A, but both normal and shear stress (force acting parallel to the surface) in the entry zone B are acting on the PLA/PVAL blends. These normal and shear stresses in the zone B affect the deformation process of the PLA droplets. As a result, the sphere droplets are rotated, elongated, and coalesced into rod-like or ellipsoid shape. Then the deformed droplets pass through the die in the zone C and may keep their shapes as they immediately leave the die exit, because the used L/D ratio of 2 of capillary die is too small and the residence time[1] of polymer blends in the die is in fact too short to deform the droplets further.

As the mass flow rate or extrusion rate increases, the normal and shear stresses acting on the PLA droplets in the entry zone B increases due to the increase of the pressure from 22.9 to 46.5 bar for mass flow rate from 0.5 to 2.0 g·min^{-1}, respectively (Figure 4.71a). Generally, according to Equation 4.9 for Newtonian system and Equation 4.10 for the specific polymer blends with the minor phase content of 15 %, the increase of shear stresses or shear rates can decrease the size of droplets. This phenomenon is often, but not always true at least in this study. For instance, the maximum CED of PLA droplets almost linearly increases from 3.6 to 12.2 μm with the increase of the mass flow rate from 0.5 to 2.0 g·min^{-1} (Figure 4.69). This unexpected phenomenon may happen in this PLA/PVAL 30/70 blends due to the coalescent

---

[1] The calculated residence times are only ca. 0.05 and 0.01 second for the mass flow rate of 0.5 and 2.0 g·min^{-1}, respectively.

process of the high content of the dispersed PLA phase and the viscoelastic behaviors of materials, which are well demonstrated by Sundararaj [240]. This anomaly has also been reported by Roland and Böhm [241], and other researchers [242, 243]. Roland and Böhm [241] concluded that the droplets have higher approach velocities at higher shear rates and therefore the coalescence probability can increase. According to Allan and Mason [244], the flow-induced coalescent mechanism of two Newtonian liquid drops can be represented as follows: A pair of particles come close to each other and rotate in the shear field, the thickness of matrix film between the particles then decreases until the interface ruptures, and coalescence occurs.

The coalescence increases with the increase of mass flow rates. This explains why the number of droplets decreases as mass flow rate increases (Figure 4.70a). It is also seen from Figure 4.69 that the average CED slightly decreases with the increase of mass flow rates. If the average CED of PLA droplets decreases, the mean length of PLA droplets must increases, because the dispersed PLA phase is under constant volume during extrusion. This result confirmed once again that the coalescence becomes more efficient at higher mass flow rates.

Comparing Figure 4.67 and Figure 4.68, it is interesting to note that the PLA droplets having CED less than ca. 0.15 μm have not been found in the remaining PLA phase after removing the PVAL matrix (Figure 4.68) but they can be seen a lot in Figure 4.67 and they occupy ca. 83, 89, 94, and 96 % of total PLA droplets for mass flow rate of 0.5, 1.0, 1.5, and 2.0 g·min^{-1}, respectively (Figure 4.70b). This observation allows one to assume that the PLA droplets having CED less than ca. 150 nm are removed together with the PVAL matrix. Therefore, it can be predicted that there has been almost no coalescence among these small droplets (the red colour droplets in Figure 4.71b) when they passing through the convergent capillary die.

Based on these above observations, the morphological development of dispersed PLA phase along the spinline should be fully studied in both cross section and longitudinal direction after etching dispersed PLA phase and removing the PVAL matrix, respectively. The mean diameters of dispersed PLA phase $(\bar{d})$[1] from PLA/PVAL blend extrudates after removing the PVAL matrix for various mass flow rates (Figure 4.72) are much larger (ca. 10 times) to that of the dispersed PLA phase in cross-sectional PLA/PVAL blend extrudates after etching PLA phase $(\bar{d}_{CED})$ (Figure 4.69) ($\bar{d} \approx 10\ \bar{d}_{CED}$). Furthermore, the mean diameter $\bar{d}$ increases with

---

[1] The mean diameter d measured from the remaining fibrils of PLA phase after removing PVAL matrix is different from circular equivalent diameter (CED) d_{CED} measured from the remaining holes of PLA phase in cross-sectional PLA/PVAL blend extrudates after etching PLA phase.

the increase of mass flow rate (Figure 4.72). This tendency is opposite to the result that was found in the cross sectional PLA/PVAL blend extrudates after etching PLA phase, that is the $\bar{d}_{CED}$ decrease with the increase of mass flow rates (Figure 4.69). This phenomenon may be caused by the two following reasons: (1) as the mass flow rate increases, the normal and shear force are acting on the droplets increases. It makes the small droplets easier to combine together to form large ones. After the Taylor's theory for the droplet deformation (Equations 2.6 – 2.10) and the capillary number (Equation 2.3), the large droplets are then deformed into ellipsoidal shapes easier than smaller ones. (2) When an ellipsoid is formed, the experimental values obtained by measurement the PLA ellipsoidal droplets on the fractured or cut surfaces may be imprecise, because one does not know where the ellipsoidal droplets are exactly cut and then measured. An ellipsoidal droplet can be cut through its center or any positions (Figure 4.73). This may explain the reason why the $\bar{d}_{CED}$ of the higher mass flow rates is slight smaller than that of the lower mass flow rates.

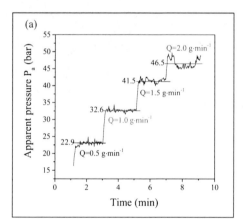 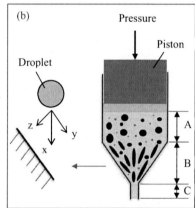

Figure 4.71 Measured pressure for various mass flow rates (a) and total pressure acting on the polymer blends in a convergent capillary die (b): the red colour droplets represent the very small droplets. There has been almost no coalescence among these small droplets when they passing though convergent capillary die.

Results and discussion

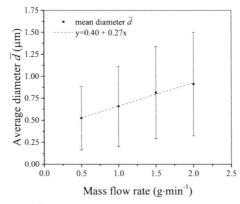

Figure 4.72 Mean diameters $\bar{d}$ with deviations of the dispersed PLA phase from PLA/PVAL blend extrudates after removing the PVAL matrix for various mass flow rates (measured from Figure 4.68).

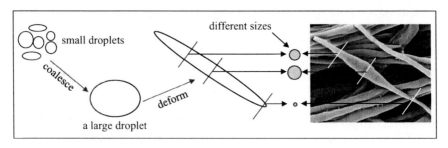

Figure 4.73 Schematic presentation of coalescence and deformation of droplets and different possible diameters of an ellipsoidal PLA droplet in PLA/PVAL blend extrudates.

Morphology of PLA/PVAL blend extrudates/filaments (P1 — P7/P8, Figure 4.55)

<u>Remaining PLA fibrils after removing PVAL matrix</u>

Figure 4.74 and 4.75 present the SEM images[1] of the PLA morphology after removing PVAL matrix from PLA/PVAL blend extrudates at different locations (P1 — P7) along the spinline for different spinning conditions. Figure 4.76 plots the mean diameter of dispersed PLA phase $\bar{d}$ versus distance to spinneret. It is obviously seen that the dispersed PLA phase is deformed from short rod-like or ellipsoidal structures in micro-scale into longer fibrillar structures in nano-scale along the spinline for all spinning conditions. For spinning condition A, in which the take-up velocity is constant, the mean diameter of dispersed PLA phase $\bar{d}$ decreases much

---

[1] Original SEM images are presented in the Appendix C.3

faster at the low mass flow rate Q=0.5 g·min⁻¹ than that of higher mass flow rates Q=1.0, 1.5, and 2.0 g·min⁻¹ (Figure 4.76a). For spinning condition B, in which the mass flow rate is constant, the mean diameter $\bar{d}$ decreases faster at the high take-up velocity v=70 m·min⁻¹ than that of lower take-up velocities v=10, 30, and 50 m·min⁻¹ (Figure 4.76b). These results clearly indicate that the above defined spinning conditions have a profound impact on the deformation of the dispersed PLA phase in PLA/PVAL blend extrudates. Under each spinning condition, the profile of filament velocity, temperature[1], tensile stress, and apparent elongational viscosity along the spinline are different as presented in the section 4.3. From the view point of the melt spinning process, the axial strain rate (ASR) (including local and maximum ASR) and the tensile stress are considered as the two most important factors that lead to the deformation of dispersed PLA phase in PLA/PVAL blend extrudates. From the view point of microrheological behavior for polymer blends in an elongational flow like melt spinning, the deformation of dispersed phase in immiscible polymer blends can also be effected by the hydrodynamic, interfacial stress, and viscosity of each polymer based on Taylor's (Equation 2.3) and Cox's theories (Equation 2.10). As mentioned in the section 2.2.4, all theories of droplet deformation and breakup are applicable for simple shear and planar flow fields with small deformations[2]. In the melt spinning, due to the complexity of non-isothermal process and very large deformations within fiber formation zone, the quantitative determination of the microrheological behavior of polymer blends is a very difficult task and almost impossible both theoretical estimation and practical measurement. For such partial miscible polymer blends like PLA/PVAL blends as discussed in the section 4.1, the determination of these parameters (interfacial stress, viscosity, etc.) becomes much more complicated due to the presence of the strong H-bonding between the two components. Therefore, the current study does not tend to quantitatively identify these microrheological parameters, but it is preferred to focus on the qualitative analysis of the deformation, breakup, and coalescence of dispersed PLA phase from empirical results of PLA/PVAL morphology. It is also preferred to concentrate on the filament parameters, with special emphasis on the two former factors: the ASR and the tensile stress of the PLA/PVAL blend extrudates within fiber formation zone that can help to understand the mechanism of deformation, break up, and coalescence of dispersed PLA phase in PLA/PVAL blend extrudates.

---

[1] Except the filament temperature profiles are nearly the same for various take-up velocities at the constant mass flow rate of 1.0 g·min⁻¹.
[2] Even the affine deformation theory of droplet deformation was also not applicable for blend extrudates within fiber formation zone in the melt spinning process, which are analyzed in the Appendix D.2

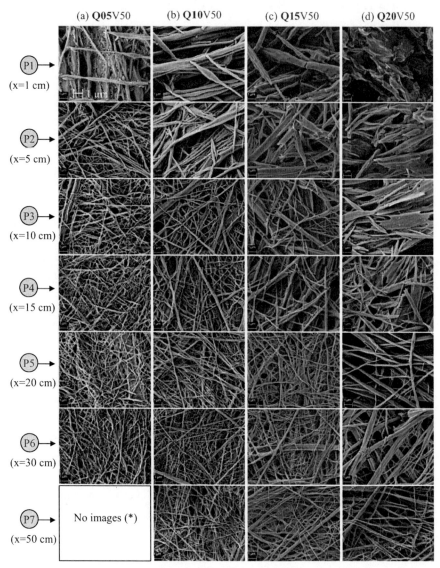

Figure 4.74 SEM images of dispersed PLA phase after removing PVAL matrix for the various mass flow rates (0.5 – 2.0 g·min^{-1}) (Q05 – Q20) and the constant take-up velocity of 50 m·min^{-1} (V50) (Q05V50, Q10V50, Q15V50, Q20V50) at different locations (P1 – P7) along the spinline: scale bar is 1 μm, (*) Experiments were not done at this location because it was supposed that there is no difference in PLA morphology at this location with that at x=30 cm (P6).

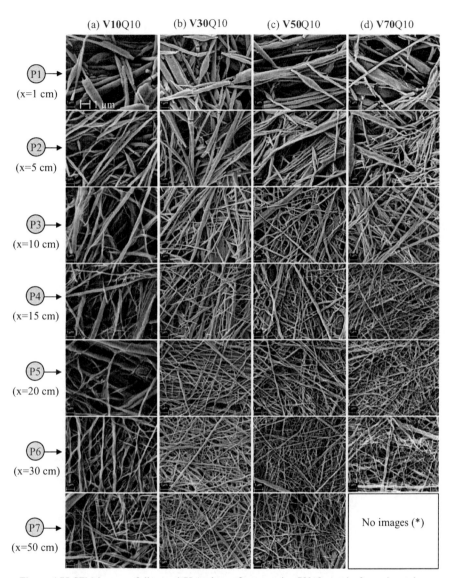

Figure 4.75 SEM images of dispersed PLA phase after removing PVAL matrix for various take-up velocities (10 – 70 m·min^{-1}) (V10 – 70) and the constant mass flow rate of 1.0 g·min^{-1} (Q10) (V10Q10, V30Q10, V50Q10, V70Q10) at different locations (P1 – P7) along the spinline: scale bar is 1 μm, (*) Experiments were not done at this location because it was supposed that there is no difference in PLA morphology at this location with that at x=30 cm (P6)

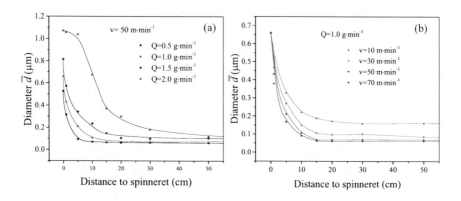

Figure 4.76 Mean diameter $\bar{d}$ of dispersed PLA phase after removing PVAL matrix vs. distance

Table 4.6 lists the maximum ASR and its locations to spinneret for different spinning conditions. Figure 4.77 plots the maximum ASR value and its position versus mass flow rate and take-up velocity. It is seen that the maximum ASR almost linearly decreases with the increase of mass flow rate at the constant take-up velocity (Figure 4.77a) and it is linearly proportional to take-up velocity for the constant mass flow rate (Figure 4.77b). Comparing these results with the SEM images in Figures 4.74 and 4.75, and with diagrams in Figure 4.77 reveals that an increase in the maximum ASR value, i.e. the decrease of mass flow rate at constant take-up velocity or the increase of take-up velocity at constant mass flow rate, cause a significant decrease in the final size of dispersed PLA phase.

Table 4.6 Maximum axial strain rate (ASR) and its locations for different spinning conditions

Conditions	Take-up velocity (m·min^{-1})	Mass flow rate (g·min^{-1})	Maximum ASR (s^{-1})	Distance to spinneret (cm)
A	50	0.5	10.61	7.5
		1.0	7.67	10
		1.5	3.03	15
		2.0	2.28	15
B	10		1.23	7.5
	30	1.0	2.59	10
	50		7.67	10
	70		9.07	10

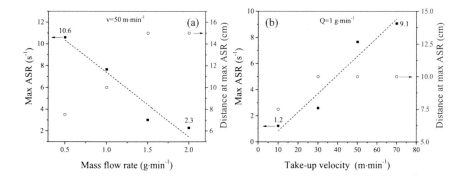

Figure 4.77 Maximum ASR and the position of maximum ASR vs. mass flow rate (a) and take-up velocity (b)

All the filament parameters at maximum ASR, which were discussed in the section 4.3, should be now reconsidered. The filament temperature and the apparent elongational viscosity at maximum ASR are represented in Figure 4.78 (can be also found in Figure 4.39, 4.40, and 4.52, section 4.3). It is seen that the filament temperature at maximum ASR goes just below melting temperature of PLA $T_{m,PLA}$ and it is much higher than its glass transition temperature. At this location, the apparent elongational viscosity has a value either equal or slightly higher than its minimum value (can be also seen in Figure 4.52c and 4.52d). Thus, the filament state at maximum ASR is under the best conditions for the filament deformation.

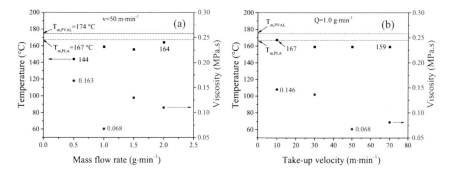

Figure 4.78 Temperature and apparent elongational viscosity of filament at maximum ASR vs. mass flow rate (a) and take-up velocity (b)

Results and discussion

It should be remembered from Table 4.5 and Figure 4.51 that tensile stress at maximum ASR decreases with the increase of the mass flow rate at the constant take-up velocity and it increases with the increase of the take-up velocity at the constant mass flow rate. This tendency is similar to that of maximum ASR as discussed above. This means that the higher the value of maximum ASR, the higher the value of tensile stress is. The simultaneous increase of both the tensile stress and maximum ASR leads to an increase in the deformation of filament. In other words, the filament deformation becomes more effective in both cases: decreasing the mass flow rate at the constant take-up velocity and increasing the take-up velocity for the constant mass flow rate.

Let us turn our attention back to the PLA morphologies in Figure 4.74 and 4.75, special in Figure 4.74d (the last column on the right of the Figure 4.74) and Figure 4.75a (the first column on the left of the Figure 4.75). These PLA morphologies were obtained under the two special spinning conditions[1]: (1) the highest mass flow rate $Q=2.0$ g·min^{-1} with the take-up velocity $v=50$ m·min^{-1}, (2) the lowest take-up velocity $v=10$ m·min^{-1} with the mass flow rate $Q=1.0$ g·min^{-1}. It is seen that the diameter of the PLA fibrils along the spinline in these two special spinning conditions decreases more slowly than that of other spinning conditions. The mean diameters of the PLA fibrils $\bar{d}$ in these spinning conditions are larger than that of other spinning conditions (Table 4.7, Figure 4.79). Furthermore, the lengths of the PLA fibrils are not endless with an average length of fibrils ca. $4-5$ μm (Figure 4.80). This could be due to little coalescence or absence of coalescence and small deformation rate. In these two special spinning conditions, the maximum ASR has the lowest values. The maximum ASR for the mass flow rate $Q=1.0$ g·min^{-1} and take-up velocity $v=10$ m·min^{-1} is only ca. 1.23 s^{-1} (Table 4.6 and Figure 4.77). Furthermore, the tensile stress at the maximum ASR has also the lowest values, which are discussed and presented in Table 4.5 and Figure 4.51.

In contrast to the two above special spinning conditions, it is seen from Figure 4.74a (the first column on the left of Figure 4.74) and 4.75d (the last column on the right of Figure 4.75) that the diameter of PLA fibrils more rapidly decreases along the spinline. These PLA fibrils were obtained under the two limiting spinning conditions[2]: (1) $Q=0.5$ g·min^{-1} and $v=50$ m·min^{-1}; (2) $Q=1.0$ g·min^{-1} and $v=70$ m·min^{-1}. The final diameters of PLA fibrils prepared using these limiting spinning conditions are much finer than that of other spinning conditions (Figure 4.79

---

[1] To distinguish the "two special spinning conditions" from the "two limiting spinning conditions"
[2] Due to the stability of the melt spinning process, mass flow rates could not be decreased less than 0.5 g·min^{-1} for a take-up velocity of 50 m·min^{-1}, and take-up velocity could not be increased more than 70 m·min^{-1} for a mass flow rate of 1.0 g·min^{-1}.

Results and discussion

and Figure 4.81), especially in comparison with the PLA fibril diameters obtained using the above special spinning conditions (Figure 4.80). In these two limiting spinning conditions, the maximum ASR and the tensile stress at maximum ASR have the highest values in comparison with other spinning conditions: A maximum ASR ranging from ca. 9.1 to 10.6 s^{-1} (Table 4.6 and Figure 4.77) and a tensile stress at maximum ASR varying from 0.7 to 1.7 MPa.s (Table 4.5 and Figure 4.51). Like the PLA fibrils prepared using the two special spinning conditions, the length of PLA fibrils at the position P6/P8 obtained using the two limiting spinning conditions appears to be also limited. However, it seems to be that these very fine fibrils are connected together at their ends to form continuous fibrils, which is seen as a nanofibrous network (Figure 4.81). It is also seen from Figure 4.81 that a few of these very fine fibrils (ca. 30 nm in diameter) could have been broken-up after reaching their maximum deformation.

Table 4.7 Average diameters $\bar{d}$ of PLA fibrils after removing PVAL matrix at locations P6 $\bar{d}_{x30}$ (x=30 cm), P7 $\bar{d}_{x50}$ (x=50 cm), and P8 $\bar{d}_L$ (x=200 cm)

Conditions	v (m·min^{-1})	Q (g·min^{-1})	$\bar{d}_{x30}$ or $\bar{d}_{x50}$ (μm)	x_{30} or x_{50} (cm)	$\bar{d}_L(x_L$=200 cm) (μm)
A	50	0.5	0.062	30	0.061
		1.0	0.068	50	0.067
		1.5	0.102	50	0.089
		2.0	0.119	50	0.092
B	10		0.160	50	0.148
	30	1.0	0.085	50	0.084
	50		0.068	50	0.067
	70		0.062	30	0.062

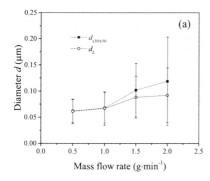
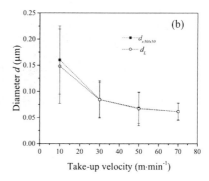

Figure 4.79 Average diameter of PLA fibrils at P6 (x=30 cm) $\bar{d}_{x30}$, P7 (x=50 cm) $\bar{d}_{x50}$, and P8 $\bar{d}_L(x_L = 200$ cm) vs. mass flow rate (a) and take-up velocity (b)

Results and discussion

Figure 4.80 SEM images of dispersed PLA phase from PLA/PVAL blend extrudates at P7 (left) and PLA/PVAL blend filaments at P8 (right) for the two special spinning conditions: Q=2.0 g·min^{-1} and v=50 m·min^{-1} (a and b), v=10 m·min^{-1} and Q=1.0 g·min^{-1} (c and d). Scale bar: 1 μm

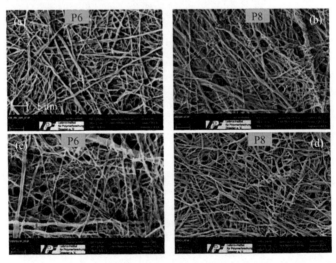

Figure 4.81 SEM images of dispersed PLA phase from PLA/PVAL blend extrudates at P6 (left) and PLA/PVAL blend filaments at P8 (right) for the two limiting spinning conditions: Q=0.5 g·min^{-1} and v=50 m·min^{-1} (a and b), v=70 m·min^{-1} and Q=1.0 g·min^{-1} (c and d). Scale bar: 1 μm

For the last three spinning conditions: (1) Q=1.0 g·min^{-1} and v=30 m·min^{-1}; (2) Q=1.0 g·min^{-1} and v=50 m·min^{-1}; (3) Q=1.5 g·min^{-1} and v=50 m·min^{-1}, in which the maximum ASR and the tensile stress at maximum ASR, respectively, vary over the range from ca. 2.6 to 7.7 s^{-1} (Table 4.6 and Figure 4.77) and from 0.35 to 0.52 MPa.s (Table 4.5 and Figure 4.51). It is seen from Figure 4.82 that the remaining PLA fibrils after removing PVAL matrix at the position P7 or P8 have also limited lengths. However, like the PLA fibrils obtained using the two limiting spinning conditions, most of the PLA fibrils seem to be joined each other at their ends to form longer fibrils, only a few very fine PLA fibrils with diameters less than ca. 50 nm are not connected together to form long continuous fibrils.

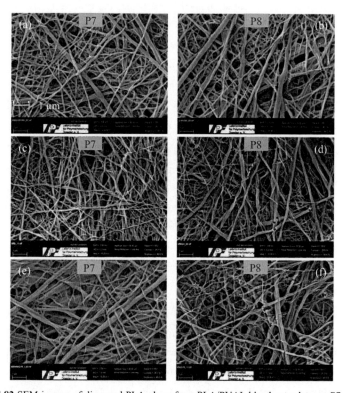

Figure 4.82 SEM images of dispersed PLA phase from PLA/PVAL blend extrudates at P7 (left) and from PLA/PVAL blend filaments at P8 (right) for the three last spinning conditions: Q=1.0 g·min^{-1} and v=30 m·min^{-1} (a and b); Q=1.0 g·min^{-1} and v=50 m·min^{-1} (c and d); Q=1.5 g·min^{-1} and v=50 m·min^{-1} (e and f). Scale bar: 1 μm

From the above analyses of the morphological development of the remaining PLA phase from PLA/PVAL blend extrudates/filaments after removing PVAL matrix and the filament profiles along the spinline, it can be said that the PLA morphology in PLA/PVAL blend extrudates can be controlled by the changes in the spinning conditions, that are the changes in mass flow rates and/or take-up velocity, i.e. the variations of the ASR, maximum ASR, and of tensile stress. Under certain spinning conditions, the short PLA fibrils are deformed and coalescence to form the long continuous fibrils.

<u>PLA morphology in the cross-sectional surfaces of PLA/PVAL blend extrudates</u>

The PLA morphologies in the cross-sectional PLA/PVAL blend extrudates after etching the dispersed PLA phase at different locations (P1 — P7) along the spinline for different spinning conditions are shown in Figure 4.83 and 4.84. Figure 4.85 plots the mean circular equivalent diameter (CED) of the dispersed PLA phase versus distance to spinneret. Generally, like the mean diameter of the dispersed PLA phase after removing PVAL matrix (abbr. the mean diameter $\bar{d}$) (Figure 4.76), the mean CED of the dispersed PLA phase after etching PLA phase (abbr. the mean CED $\bar{d}_{CED}$) (Figure 4.85) decreases along the spinline for all spinning conditions. However, in some cases, it is seen from Figure 4.85 that the mean CED $\bar{d}_{CED}$ along the spinline slightly becomes larger after it initially decreases. The exact cause of this phenomenon is not known: It could be due to the radial coalescence of the neighbour PLA droplets through cross-section of PLA/PVAL blend extrudates during stretching process as schematically shown in Figure 4.86. Or it could be preferable due to the imperfect fractured surfaces of the cross-sectional PLA/PVAL blend extrudates and not uniform of fractured positions of PLA/PVAL blend extrudates as discussed in Figure 4.73. The latter reason is more acceptable than the former reason, because this phenomenon was not found for the dispersed PLA phase after removing PVAL matrix (Figure 4.67).

Comparing Figures 4.74 and 4.83, Figures 4.75 and 4.84, and Figures 4.76 and 4.85, an important observation can be found that many very fine PLA droplets/fibrils on fractured surfaces of the cross-sectional PLA/PVAL blend extrudates after etching PLA phase do not appear in the remaining PLA phase after removing the PVAL matrix. Furthermore, the mean CEDs are always smaller than the mean diameters along the spinline ($\bar{d}_{CED} < \bar{d}$). The similar results have already been found for the PLA/PVAL blend extrudates without stretching in the above section "Morphology of PLA/PVAL blend extrudates". To make this result clearer, the three pairs of the SEM images of PLA/PVAL blend extrudates (Figure 4.87) at the position P6 (x=30 cm) after removing the PVAL matrix (images on the right column) and etching the dispersed PLA phase (images on the left column) from the three last different spinning conditions[1] are selected to analyze the diameter d and CED d_{CED} of PLA fibrils. Figure 4.88 gives the mean diameter $\bar{d}$ and the mean CED $\bar{d}_{CED}$ of dispersed PLA fibrils. It is seen that the mean diameters are always larger than the mean CEDs for all the three selected spinning conditions $\bar{d} > \bar{d}_{CED}$. Furthermore, it is also obviously seen from Figure 4.89 that the cumulative number percentage of PLA droplets having diameter up to 0.1 µm was always more than that of PLA droplets having CED up to 0.1 µm. Especially for the spinning condition with Q=1.0 g·min^{-1} and v=30 m·min^{-1}, the difference of cumulative number percentage of PLA droplets between the diameter d and CED d_{CED} becomes larger: While there are ca. 72.1 % the number of PLA droplets having CED up to 0.1 µm, there are up to 90 % the number of PLA droplets having diameter up to 0.1 µm. These results may allow one to confirm once again that most of very fine PLA droplets are removed together with the PVAL matrix during removing process due to less or almost no coalescence among these very fine droplets or between them and other larger neighbour droplets.

<u>Possible conceptual models of the fibrillation process of PLA/PVAL blend extrudates</u>

Based on all the above analyses of the morphological development of the dispersed PLA phase and the profiles of PLA/PVAL blend filament along the spinline, an overview of possible conceptual models of the deformation, coalescence, and break-up processes of the dispersed PLA droplets in PLA/PVAL blend extrudates during melt spinning within fiber formation zone is summarized in Table 4.8 and schematically visualized in more details in Figure 4.90. Figure 4.91 describes the deformation levels for different spinning conditions. The possible conceptual models for the fibrillation process can occur in the following sequences, which are also summarized in Table 4.9: the dispersed PLA droplets/fibrils are either deformed and no coalesced; or deformed and coalesced and further deformed; or deformed and coalesced and further deformed and broken-up along the spinline. From the view point of the melt spinning process, the fibrillation process depends on the ASR and tensile stress level. Generally, like other polymer processes such as mixing and compounding, it strongly depends on the droplet size distribution, the blend ratio, and the microrheological behavior of blends.

---

[1] Under these spinning conditions, the PLA droplets/fibrils seem to be more coalescent than other spinning conditions

Results and discussion

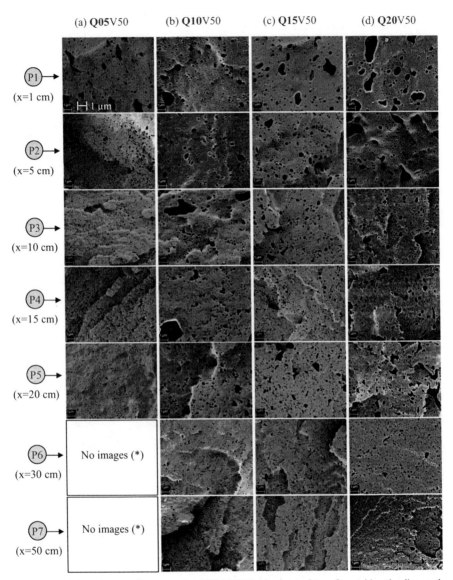

Figure 4.83 SEM images of cross-sectional PLA/PVAL blend extrudates after etching the dispersed PLA phase for the various mass flow rates (0.5 – 2.0 g·min^{-1}) (Q05 – Q20) and the constant take-up velocity of 50 m·min^{-1} (V50) (Q05V50, Q10V50, Q15V50, Q20V50) at different locations (P1 – P7) along the spinline: scale bar 1 μm, (*) Experiments were not done at this location because it was supposed that there is no difference in PLA morphology at this location with that at x=20 cm (P5)

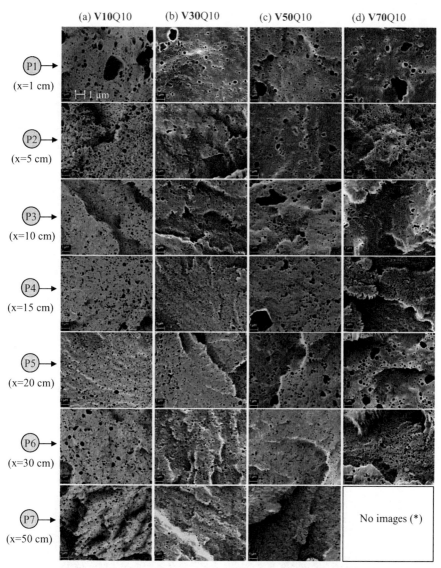

Figure 4.84 SEM images of cross-sectional PLA/PVAL blend extrudates after etching the dispersed PLA phase for various take-up velocities (10 – 70 m·min^{-1}) (V10 – V70) and the constant mass flow rate of 1.0 g·min^{-1} (Q10) (V10Q10, V30Q10, V50Q10, V70Q10) at different locations (P1 – P7) along the spinline: scale bar 1 µm, (*) Experiments were not done at this location because it was supposed that there is no difference in PLA morphology at this location with that at x=30 cm (P6)

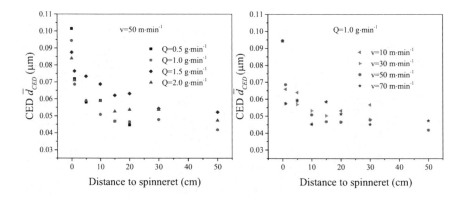

Figure 4.85 Mean CED of the dispersed PLA phase in cross-sectional PLA/PVAL blend extrudates after etching the PLA phase vs. distance to spinneret

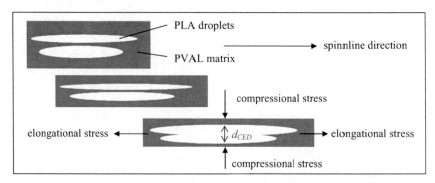

Figure 4.86 A schematic draw of a possible radial coalescence of neighbour PLA droplets in a PLA/PVAL blend extrudate during stretching process under effect of elongational and compressional stresses

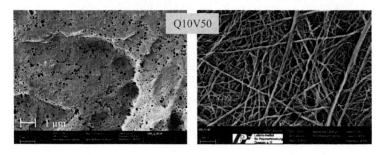

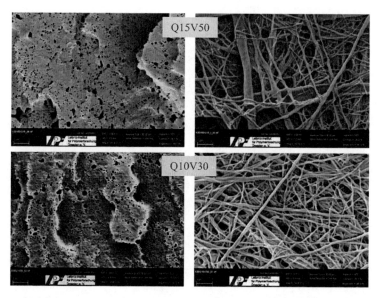

Figure 4.87 SEM images of PLA/PVAL blend extrudates at position P6 (x=30 cm) after removing PVAL matrix (right column) and etching dispersed PLA phase (left column) for the three different spinning conditions: Q=1.0 g·min^{-1} and v=50 m·min^{-1} (Q10V50); Q=1.5 g·min^{-1} and v=50 m·min^{-1} (Q15V50); Q=1.0 g·min^{-1} and v=30 m·min^{-1} (Q10V30), scale bar: 1 µm

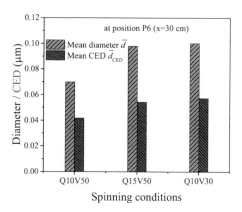

Figure 4.88 Mean diameter $\bar{d}$ and mean CED $\bar{d}_{CED}$ of the dispersed PLA phase from PLA/PVAL blend extrudates at position P6 (x=30 cm) for the three different spinning conditions: Q=1.0 g·min^{-1} and v=50 m·min^{-1} (Q10V50); Q=1.5 g·min^{-1} and v=50 m·min^{-1} (Q15V50); Q=1.0 g·min^{-1} and v=30 m·min^{-1} (Q10V30)

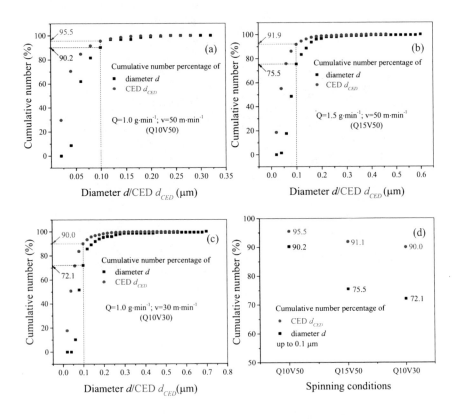

Figure 4.89 Cumulative number percentage vs. diameter d/CED d_{CED} of the dispersed PLA phase from PLA/PVAL blend extrudates at position P6 (x=30 cm) for the last three different spinning conditions: (a) Q=1.0 g·min^{-1} and v=50 m·min^{-1} (Q10V50); (b) Q=1.5 g·min^{-1} and v=50 m·min^{-1} (Q15V50); (c) Q=1.0 g·min^{-1} and v=30 m·min^{-1} (Q10V30). (d) Cumulative number percentage of diameter d/CED d_{CED} having diameter up to 0.1 µm

Table 4.8 An overview of the possibility of the deformation (De), coalescence (Co), and break-up (Bu) processes for different spinning conditions

Spinning conditions (SCD)		
Group I (Two specific SCD)	Group II (Three last/middle SCD)	Group III (Two limiting SCD)
$Q=1.0$ g·min^{-1}; $v=10$ m·min^{-1}	$Q=1.0$ g·min^{-1}; $v=50$ m·min^{-1}	$Q=0.5$ g·min^{-1}; $v=50$ m·min^{-1}
$Q=2.0$ g·min^{-1}; $v=50$ m·min^{-1}	$Q=1.0$ g·min^{-1}; $v=30$ m·min^{-1}	$Q=1.0$ g·min^{-1}; $v=70$ m·min^{-1}
	$Q=1.5$ g·min^{-1}; $v=50$ m·min^{-1}	
Filament parameters		
Low max. ASR (1.2 – 2.3 s^{-1}) and low tensile stress	Middle max. ASR (2.6 – 7.7 s^{-1}) and middle tensile stress	High max. ASR (9.1 – 10.6 s^{-1}) and high tensile stress
The possibility of deformation, coalescence, and break-up		
Less deformation	More deformation	The most deformation
Less axial coalescence[1]	More axial coalescence	The most axial coalescence
No fibril break-up[2]	Almost no break-up	More break-up

[1] The coalescence in the direction along the spinline
[2] Break-up for only the fibrils, which are formed after deformation and coalescence, not for the droplets

Results and discussion

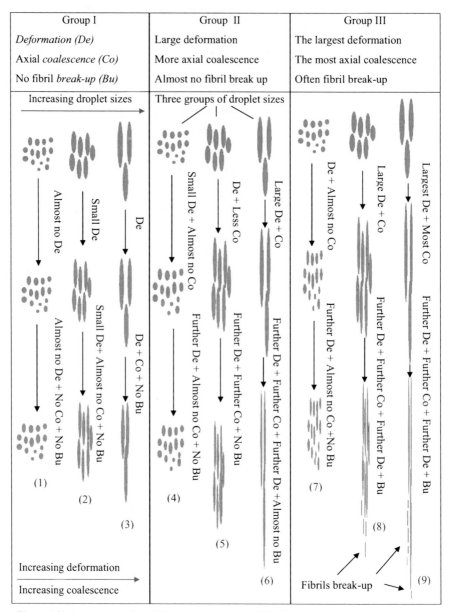

Figure 4.90 An overview of possible conceptual models of the deformation, coalescence, and break-up processes for different spinning conditions

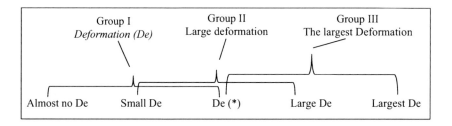

Figure 4.91 A relative scale of the deformation levels. (*) The term "deformation (De)" does not mean that its deformation level in Group I is the same value as its deformation level in Group II or III. All the terms "Almost no De", "Small De", "De", "Large De", and "Largest De" are used to compare the deformation level within each group.

Table 4.9 Different possible sequences of droplet deformation, coalescence, and break-up

Sequences of deformation (De), coalescence (Co), and break-up (Bu)	Sequences number
- De but no Co and no Bu	(1)
- De and Co, but no Bu	(2) and (3)
- De and Co, then further De and Co, but no Bu	(4), (5), and (7)
- De and Co, then further De and Co, and then further De and Bu	(6), (8), and (9)

Chapter 5
Conclusions and future works

5.1 Conclusions

A novel and simple fabrication process for producing biodegradable and biocompatible nanofibrillar poly(lactic acid) structures from poly(lactic acid) (PLA) and poly(vinyl alcohol) (PVAL) blends has been successfully developed by using the conventional melt spinning method (CMS) of polymeric fibers on the industrial-scale melt spinning device with twin-screw extruder. The melt spinning process temperatures were optimized by determining the best combinations of the thermal behavior and rheological properties of PLA/PVAL blends. It was found that the optimal extruder temperature profile was 175, 180, 185, 190, and 195 °C (starting from the feeding zone to the die). Thus, for the first time, different non-commercial PLA/PVAL blend filament yarns (including the nanofibrillar PLA structures) with different finenesses were fabricated under various take-up velocities ranging from 30 − 50 m·min^{-1}. It was found that the melt spinning process at the take-up velocity of 30 m·min^{-1} is highly reproducible and the most stable process. The textile-physical properties of PLA/PVAL blend filaments were also found to be optimum at the take-up velocity of 30 m·min^{-1}. These filaments were then drawn with different draw ratios (DR). The off-line drawn PLA/PVAL filaments with the draw ratio of 1.70 provided the highest tenacity. The textile-physical properties of these off-line drawn filaments with DR=1.70 are similar to that of acetate rayon or wool fibers. It was found that the produced as-spun and drawn PLA/PVAL filaments have high enough strength and suitable elongation for further processing in various textile processes such as weaving and knitting. Thus, the production form "PLA/PVAL filament yarns" of the CMS provides great flexibility in producing and designing nanofibrillar PLA structures or nanofibrous PLA scaffolds, which are made from woven and knitted fabrics of PLA/PVAL filament yarns after removing PVAL matrix.

Conclusions and future works

The main benefit in using the CMS is its very high productivity. Other advantages of the CMS are the solvent free process and low energy consumption. Owing to these advantages of the CMS with the combinations of the water solubility, non-toxicity, biodegradability, and biocompatibility of PVAL, which is used as a matrix component in PLA/PVAL blends, the investigated fabrication process provided great potential to produce the biodegradable and biocompatible nanofibrous PLA scaffolds without any additional organic solvents. Moreover, the produced PLA fibrils after removing PVAL matrix from PLA/PVAL blend filaments were in nanometer scale range with average diameter of PLA fibrils varying from 44 to 85 nm depending on the take-up velocity and the draw ratio. Thus, nanoporous PLA scaffolds made from PLA/PVAL blend filaments could be promising candidates as scaffolds for tissue engineering due to their outstanding biocompatibility and a large-surface-area-to-volume ratio. Although it is too early to provide a particular application for nanofibrillar PLA structures, presentative nanofibrous scaffolds for bone and skin tissue engineering are proposed by considering the size of the PLA fibrils, the three dimensional interconnected networks, and the mechanical properties of nanofibrillar PLA scaffolds.

Besides the development and fabrication of the desired nanofibrillar PLA structures using the CMS on the industrial-scale melt spinning device, the present research study focused its attention on the characterization of the PLA/PVAL monofilament profiles along the spinline in the low-speed melt spinning process under various spinning conditions. Thanks to the combination of the filament velocity and the filament temperature measurements using the laser Doppler velocimetry and the infrared thermography, respectively, the fiber formation zone was determined. It is interesting to note that the length of the fiber formation zone according to the filament velocity profiles L_{Sv} is not the same as the length of the fiber formation zone according to the filament temperature profiles L_{ST} ($L_{ST} > L_{Sv}$). This unexpected phenomenon may happen for the low-speed melt spinning process due to the axial heat conduction effect of the filament along the spinline and the non-uniform radial temperature through the cross-sectional thick filament. Based on the experimental results of the filament temperature profiles, it was found that the Nusselt number is nearly constant and has values 0.85 and 0.90. Special attention of the characterization activity of filament profiles was given to the maximum axial strain rate (ASR) and tensile stress at maximum ASR, which are recognized as two important parameters that influence the final state of deformation of the dispersed PLA phase in PLA/PVAL filaments. It was demonstrated that the maximum ASR and tensile stress at maximum ASR decrease with the increase of the mass flow rate at the constant take-up velocity and they both increase with increasing of take-up velocity at the constant mass flow rate.

The most challenging task of the present research activity was the understanding of the micro- and nanofibrillation process of PLA/PVAL blend filaments along the spinline. Owing to a new special self-constructed fiber capturing device at IPF Dresden, valuable insight was gained into the morphology development of the PLA/PVAL blend extrudates within fiber formation zone under various spinning conditions for the first time. It was found that fibrillation process of the dispersed PLA phase from the rod-like to nanofibrillar structures mainly occurs in the fiber formation zone under the effect of an elongational flow. In this zone, the PLA/PVAL blend was converted from the molten to viscoelastic state; the filament velocity and tensile stress increased rapidly. It was shown that the final sizes of PLA fibrils can be controlled by changes in the spinning conditions via the take-up velocity and flow rate. The final diameter of PLA fibrils become finer if the mass flow rate decreases for the constant take-up velocity or the take-up velocity increases for the constant flow rate. The spinning conditions could be roughly classified into three groups, depending on the value of maximum axial strain rate (ASR) and the tensile stress. It was shown that the higher the value of maximum ASR and tensile stress, the more PLA fibrils deformed and coalesced. However, if the maximum ASR and tensile stress at maximum ASR went to their upper limiting values of 9 s^{-1} and 0.74 MPa, respectively, the break-up of deformed and coalesced PLA fibrils with a very fine diameter (ca. 30 nm) took-place. On the contrary, if the maximum ASR and tensile stress at maximum ASR had low values of 1.23 s^{-1} and 0.18 MPa, respectively, the coalescence process almost did not occur. The lengths of PLA fibrils were not endless with an average length of ca. 4 − 5 μm. It is suggested that the optimal spinning conditions for the formation of long continuous PLA fibrils in PLA/PVAL blends should be customized with the maximum ASR range from ca. 3 to 9 s^{-1}. Another major contribution of the present work was that the analysis of the morphological development of the dispersed PLA phase in PLA/PVAL blend extrudates along the spinline were done in the both cross-sectional and longitudinal direction of PLA/PVAL blend extrudates: the PLA fibrils after removing the PVAL matrix and the remaining holes of PLA fibrils in cross-sectional PLA/PVAL blend extrudates after etching the PLA phase were simultaneously studied. The results demonstrated that there is almost no axial coalescence between small PLA droplets/fibrils during stretching of PLA/PVAL blend extrudates within fiber formation zone. In contrast, the big droplets /fibrils are well deformed and coalesced in an elongational flow under the effect of the ASR

Conclusions and future works

and the tensile stress. The deformed and coalesced fibrils of the above mentioned big droplets/fibrils are further deformed. They reach to a lower critical diameter of ca. 30 nm. And they will then break-up into short thin fibrils if the maximum ASR is higher than its critical value in combination with the high tensile stress. Finally, possible conceptual models for the fibrillation process of the dispersed PLA phase were proposed depending on the spinning conditions and the droplet sizes.

5.2 Future works

Besides the achievements, the present study results also raised new points. I would like to briefly highlight some potential future works as follows:

- The nanofibrillar structures or nanofibrillar scaffolds from the produced PLA/PVAL filaments need to evaluate their properties by experts from other research fields related to tissue engineering or biomedical materials to find suitable applications before they can be transformed into commercial products.
- The present findings confirmed that a very small amount of PVAL still remaining in PLA fibrils due to the partial miscibility of PLA/PVAL blends. This definitely affects the PLA morphology and the mechanical properties of PLA nanofibrils/nanoscaffolds made from PLA nanofibrils. For instance, the PLA nanofibrils and nanoscaffolds were found to have a low elongation. The improvement of the mechanical properties of PLA nanofibrils/ nanoscaffolds should be an important and a challenging future task.
- In the extremely low to low melt spinning process of PLA/PVAL blends, it was found that there is the difference between solidification lengths from velocity profiles and from temperature profiles. The exact cause of this unexpected phenomenon is not known: Is it caused either by the axial heat conduction effect, by the non-uniform radial temperature through cross-sectional filament, or by the combination of both the factors? The answer to these open questions is relevant because it can help to understand the morphology development of polymer blends in the two- and three-dimensional models of the melt spinning process.

References

1. Ko F. K. and Gandhi M. R., in *Nanofibers and nanotechnology in textiles*, Brown P.J., Stevens K., and Stevens K.A., Eds, Woodhead Publishing: Cambridge, England **2007**.
2. Zhang Y., Lim C., Ramakrishna S., and Huang Z.-M., Recent development of polymer nanofibers for biomedical and biotechnological applications. Journal of Materials Science: Materials in Medicine 16(10), 933-946 **(2005)**
3. Hongu T., Phillips G.O., and Takigami M., in *New Millennium Fibers*, Woodhead Publishing: Boca Raton, Boston, New York, Washington DC **2005**.
4. Shirai H. and Yamaura K., in *"Fiber Kougaku"*: Tokyo, Japan **2005**.
5. Nakata K., et al., Poly(ethylene terephthalate) Nanofibers Made by Sea–Island–Type Conjugated Melt Spinning and Laser-Heated Flow Drawing. Macromolecular Rapid Communications 28(6), 792-795 **(2007)**
6. Ohkoshi Y., in *Handbook of Textile Fibre Structure: Fundamentals and Manufactured Polymer Fibres*, Eichhorn S., Hearle J.W.S., and Kikutani T., Eds, Woodhead Publishing: Oxford, Cambridge, New Delhi **2009**.
7. Hersh S. P., in *Handbook of Fiber Science and Technology Volume 3: High Technology Fibers*, Lewin M. and Preston J., Eds, Marcel Dekker, Inc: New York, Basel **1985**.
8. Ko F.K. and Wan Y., in *Introduction to Nanofiber Materials*, Ko F.K. and Wan Y., Eds, Cambridge University Press: Cambridge, England **2014**.
9. Shitov N. A., Timofeeva G. I., and Aizenshtein É M., Preparation of ultra-thin fibres from polymer mixtures. Fibre Chemistry 17(5), 305-311 **(1986)**
10. Kamath M. G. and Bhat G. S., in *Polyesters and Polyamides*, Deopura B. L., et al., Eds, Woodhead Publishing: Cambridge England **2008**.
11. Paul D. R., in *Polymer Blends*, Paul D. R. and Newman S., Eds, Academic Press **1978**.
12. Elias H.-G., in *Macromolecules Volume 2. Synthesis, Materials, and Technology*, Elias H.-G., Editor, Springer US: New York **1984**.
13. Tsebrenko M. V., Jakob M., Kuchinka M. Y., Yudin A. V., and Vinogradov G. V., Fibrillation of Crystallizable Polymers in Flow Exemplified by Melts of Mixtures of Polyoxymethylene and Copolyamides. International Journal of Polymeric Materials and Polymeric Biomaterials 3(2), 99-116 **(1974)**
14. Tsebrenko M. V., Fibrillation of the Mixtures of Crystallizable, Amorphous and Poorly Crystalline Polymers. International Journal of Polymeric Materials and Polymeric Biomaterials 10(2), 83-119 **(1983)**
15. Evstatiev M., Fakirov S., and Friedrich K., Effect of blend composition on the morphology and mechanical properties of microfibrillar composites. Applied Composite Materials 2(2), 93-106 **(1995)**

References

16. Fakirov S., Bhattacharyya D., Lin R. J. T., Fuchs C., and Friedrich K., Contribution of Coalescence to Microfibril Formation in Polymer Blends during Cold Drawing. Journal of Macromolecular Science, Part B **46**(1), 183-194 **(2007)**

17. Jayanarayanan K., Jose T., Thomas S., and Joseph K., Effect of draw ratio on the microstructure, thermal, tensile and dynamic rheological properties of insitu microfibrillar composites. European Polymer Journal **45**(6), 1738-1747 **(2009)**

18. Li W., Schlarb A.K., and Evstatiev M., Influence of processing window and weight ratio on the morphology of the extruded and drawn PET/PP blends. Polymer Engineering & Science **49**(10), 1929-1936 **(2009)**

19. Yi X., et al., Morphology and properties of isotactic polypropylene/poly(ethylene terephthalate) in situ microfibrillar reinforced blends: Influence of viscosity ratio. European Polymer Journal **46**(4), 719-730 **(2010)**

20. Miller W.A. and Merriam J.C.N., *Synthetic fibrous products*, Patent No. 3097991, US. **1963**.

21. Min K., White J. L., and Fellers J. F., High density polyethylene/polystyrene blends: Phase distribution morphology, rheological measurements, extrusion, and melt spinning behavior. Journal of Applied Polymer Science **29**(6), 2117-2142 **(1984)**

22. Liang B.-R., White J. L., Spruiell J. E., and Goswami B. C., Polypropylene/nylon 6 blends: Phase distribution morphology, rheological measurements, and structure development in melt spinning. Journal of Applied Polymer Science **28**(6), 2011-2032 **(1983)**

23. Li X., Chen M., Huang Y., and Cong G., In-situ generation of polyamide-6 fibrils in polypropylene processed with a single screw extruder. Polymer Engineering & Science **39**(5), 881-886 **(1999)**

24. Afshari M., Kotek R., Haghighat K. M., Nazock D. H., and B.S. Gupta, Effect of blend ratio on bulk properties and matrix–fibril morphology of polypropylene/nylon 6 polyblend fibers. Polymer **43**(4), 1331-1341 **(2002)**

25. Monticciolo A., Cassagnau P., and Michel A., Fibrillar morphology development of PE/PBT blends: Rheology and solvent permeability. Polymer Engineering & Science **38**(11), 1882-1889 **(1998)**

26. Tavanaie M. A., Shoushtari A. M., Goharpey F., and Mojtahedi M. R., Matrix-fibril morphology development of polypropylene/poly(butylenes terephthalate) blend fibers at different zones of melt spinning process and its relation to mechanical properties. Fibers and Polymers **14**(3), 396-404 **(2013)**

27. Breen A.L., *Process for preparing oriented microfibers*, Patent No. 3382305, Du Pont, US. **1968**.

28. Lyoo W. S., Choi Y. G., Choi J. H., Ha W. S., and Kim B. C., Rheological and Morphological Properties of Immiscible Blends and Microfiber Preparation from the Blends. International Polymer Processing **15**(4), 369-379 **(2000)**

29. Wang D. and Sun G., Formation and morphology of cellulose acetate butyrate (CAB)/polyolefin and CAB/polyester in situ microfibrillar and lamellar hybrid blends. European Polymer Journal **43**(8), 3587-3596 **(2007)**

30. Robeson L. M., Axelrod R. J., Vratsanos M. S., and Kittek M. R., Microfiber formation: Immiscible polymer blends involving thermoplastic poly(vinyl alcohol) as an extractable matrix. Journal of Applied Polymer Science **52**(13), 1837-1846 **(1994)**

31. Tsebrenko M. V., Rezanova N. M., and Tsebrenko I. A., Fiber-forming properties of polymer mixture melts and properties of fibers on their basis. Polymer Engineering & Science **39**(12), 2395-2402 **(1999)**

32. Fakirov S., in *Synthetic Polymer-Polymer Composites*, Bhattacharyya D. and Fakirov S., Eds, Carl Hanser Verlag GmbH & Co. KG **2012**.

33. Lin S. T. C., Bhattacharyya D., Fakirov S., and Cornish J., Novel Organic Solvent Free Micro-/Nanofibrillar, Nanoporous Scaffolds for Tissue Engineering. International Journal of Polymeric Materials and Polymeric Biomaterials **63**(8), 416-423 **(2014)**

34. Bhattacharyya D. and Fakirov S., in *Nano- and Micromechanics of Polymer Blends and Composites*, Karger-Kocsis J. and Fakirov S., Eds, Carl Hanser Verlag: Munich, Germany **2009**.

35. Wiedemann P., *Mikrofibrilläre Fasern auf Basis bioabbaubarer Polymerblends - Herstellung und Charakterisierung*, Dipl.-Ing. Thesis,Technische Universität Dresden, Germany **2011**.

References

36. Tran N. H. A., Brünig H., Boldt R., and Heinrich G., Melt Spinning of Biodegradable Nanofibrillary Structures from Poly(lactic acid) and Poly(vinyl alcohol) Blends. Macromolecular Materials and Engineering **299**(2), 219-227 **(2014)**

37. Pierce C.N., *Melt spinning apparatus and method*, Patent No. 3437725, Du Pont, US. **1969**.

38. Kwon Y.D., Kavesh S., and Prevorsek D.C., *Method of preparing high strength and modulus polyvinyl alcohol fibers*, Patent No. 4,440,711, Allied Corporation. **1984**.

39. Tanaka H., Suzuki M., and Ueda F., *Ultra-high-tenacity polyvinyl alcohol fiber and process for producing same*, Patent No. 4603083, Toray Industries, Inc. **1986**.

40. Han C.D. and Kim Y. W., Dispersed Two-Phase Flow of Viscoelastic Polymeric Melts in a Circular Tube. Transactions of The Society of Rheology (1957-1977) **19**(2), 245-269 **(1975)**

41. Tsebrenko M.V., Yudin A.V., Ablazova T.I., and Vinogradov G.V., Mechanism of fibrillation in the flow of molten polymer mixtures. Polymer **17**(9), 831-834 **(1976)**

42. Vinogradov G. V., et al., Effect of Rheological Properties of Compounds on Fiber Formation in Mixtures of Incompatible Polymers. International Journal of Polymeric Materials and Polymeric Biomaterials **9**(3-4), 187-200 **(1982)**

43. Dreval V. E., et al., Deformation of melts of mixtures of incompatible polymers in a uniform shear field and the process of their fibrillation. Rheologica Acta **22**(1), 102-107 **(1983)**

44. Min K., White J. L., and Fellers J. F., Development of phase morphology in incompatible polymer blends during mixing and its variation in extrusion. Polymer Engineering & Science **24**(17), 1327-1336 **(1984)**

45. La Mantia F. P., Valenza A., Paci M., and Magagnini P. L., Rheology-morphology relationships in nylon 6/liquid-crystalline polymer blends. Polymer Engineering & Science **30**(1), 7-12 **(1990)**

46. Favis B. D. and Therrien D., Factors influencing structure formation and phase size in an immiscible polymer blend of polycarbonate and polypropylene prepared by twin-screw extrusion. Polymer **32**(8), 1474-1481 **(1991)**

47. Bordereau V., Carrega M., Shi Z. H., Utracki L. A., and Sammut P., Development of polymer blend morphology during compounding in a twin-screw extruder. Part III: Experimental procedure and preliminary results. Polymer Engineering & Science **32**(24), 1846-1856 **(1992)**

48. Gonzalez-Nunez R., Favis B. D., Carreau P. J., and Lavallée C., Factors influencing the formation of elongated morphologies in immiscible polymer blends during melt processing. Polymer Engineering & Science **33**(13), 851-859 **(1993)**

49. Chapleau N. and Favis B. D., Droplet/fibre transitions in immiscible polymer blends generated during melt processing. Journal of Materials Science **30**(1), 142-150 **(1995)**

50. Lee J. K. and Han C. D., Evolution of polymer blend morphology during compounding in a twin-screw extruder. Polymer **41**(5), 1799-1815 **(2000)**

51. Covas J. A., Carneiro O. S., and Maia J. M., Monitoring the Evolution of Morphology of Polymer Blends Upon Manufacturing of Microfibrillar Reinforced Composites. International Journal of Polymeric Materials and Polymeric Biomaterials **50**(3-4), 445-467 **(2001)**

52. Pesneau I., Kadi A. A., Bousmina M., Cassagnau P., and Michel A., From polymer blends to in situ polymer/polymer composites: Morphology control and mechanical properties. Polymer Engineering & Science **42**(10), 1990-2004 **(2002)**

53. Filipe S., Cidade M. T., Wilhelm M., and Maia J. M., Evolution of morphological and rheological properties along the extruder length for blends of a commercial liquid crystalline polymer and polypropylene. Polymer **45**(7), 2367-2380 **(2004)**

54. Wang D., Sun G., and Chiou B.-S., Fabrication of Tunable Submicro- or Nano-Structured Polyethylene Materials from Immiscible Blends with Cellulose Acetate Butyrate. Macromolecular Materials and Engineering **293**(8), 657-665 **(2008)**

55. Padsalgikar A. D. and Ellison M. S., Modeling droplet deformation in melt spinning of polymer blends. Polymer Engineering & Science **37**(6), 994-1002 **(1997)**

References

56. Yang J., White J. L., and Jiang Q., Phase morphology development in a low interfacial tension immiscible polyolefin blend during die extrusion and melt spinning. Polymer Engineering & Science **50**(10), 1969-1977 **(2010)**

57. Hayashi T., Biodegradable polymers for biomedical uses. Progress in Polymer Science **19**(4), 663-702 **(1994)**

58. Chandra R. and Rustgi R., Biodegradable polymers. Progress in Polymer Science **23**(7), 1273-1335 **(1998)**

59. Fambri L., Migliaresi C., and Piskin E., in *Integrated Biomaterials Science*, R. Barbucci, Editor, Kluwer Academic/Plenum Publishers: New York, Boston, Dordrecht, London, Moscow **2002**.

60. Lenz R. W., in *Biopolymers I*, Langer R. S. and Peppas N. A., Eds, Springer Berlin Heidelberg **1993**.

61. Shuai X., He Y., Asakawa N., and Inoue Y., Miscibility and phase structure of binary blends of poly(L-lactide) and poly(vinyl alcohol). Journal of Applied Polymer Science **81**(3), 762-772 **(2001)**

62. Nair L. S. and Laurencin C. T., Biodegradable polymers as biomaterials. Progress in Polymer Science **32**(8–9), 762-798 **(2007)**

63. Sin L. T., Rahmat A. R., and Rahman W. A. W. A., in *Polylactic Acid PLA Biopolymer Technology and Applications*, Sin L. T., Rahmat A. R., and Rahman W. A. W. A., Eds, Elsevier Inc.: Amsterdam, Boston, Heidelberg, London, New York, Oxford, Paris, Sandiego, Sanfrancisco, Singapore, Sydney, Tokyo **2012**.

64. Haehnel W. and Herrmann W.O., *Verfahren zur Darstellung von polymerem Vinylalkohol*, Patent No. DE 450 286. **1927**.

65. Herrmann W.O. and Haehne W., *Über den Poly-vinylalkohol*. 1927, Berichte der deutschen chemischen Gesellschaft. p. 1658-1663.

66. Martin K. L., Vinyl acetate and the textile industry. Textile Chemist & Colorist **21**(1), 21-28 **(1989)**

67. Goodship V. and Jacobs D., *Polyvinyl Alcohol: Materials, Processing and Applications*, Smithers Rapra Technology **2005**.

68. Kuraray Co. Ltd. 2013; Available from: http://www.kuraray.co.jp/en/release/2013/131120.html.

69. Okaya T., in *Polyvinyl Alcohol-Developments*, C.A. Finch, Editor, John Wiley & Sons, Inc.: Chichester, New york, Brisbane, Toronto, Singapore **1992**.

70. Hu X., Mamoto R., Shimomura Y., Kimbara K., and Kawai F., Cell surface structure enhancing uptake of polyvinyl alcohol (PVA) is induced by PVA in the PVA-utilizing Sphingopyxis sp. strain 113P3. Archives of Microbiology **188**(3), 235-241 **(2007)**

71. López O. B. L., Sierra G. L., and Mejía G. A. I., Biodegradability of poly(vinyl alcohol). Polymer Engineering & Science **39**(8), 1346-1352 **(1999)**

72. Baker M. I., Walsh S. P., Schwartz Z., and Boyan B. D., A review of polyvinyl alcohol and its uses in cartilage and orthopedic applications. Journal of Biomedical Materials Research Part B: Applied Biomaterials **100B**(5), 1451-1457 **(2012)**

73. Hassan C. M. and Peppas N. A., in *Biopolymers·PVA Hydrogels, Anionic Polymerisation Nanocomposites*, Springer Berlin Heidelberg **2000**.

74. Suzuki T., Ichihara Y., Yamada M., and TonomuraKenzo., Some Characteristics of Pseudomonas O-3 which Utilizes Polyvinyl Alcohol. Agricultural and Biological Chemistry **37**(4), 747-756 **(1973)**

75. Watanabe Y., Hamada N., Morita M., and Tsujisaka Y., Purification and properties of a polyvinyl alcohol-degrading enzyme produced by a strain of Pseudomonas. Archives of Biochemistry and Biophysics **174**(2), 575-581 **(1976)**

76. Sakazawa C., Shimao M., Taniguchi Y., and Kato N., Symbiotic Utilization of Polyvinyl Alcohol by Mixed Cultures. Applied and Environmental Microbiology **41**(1), 261-267 **(1981)**

77. Shimao M., Saimoto H., Kato N., and Sakazawa C., Properties and roles of bacterial symbionts of polyvinyl alcohol-utilizing mixed cultures. Applied and Environmental Microbiology **46**(3), 605-610 **(1983)**

78. Kim M. N. and Yoon M. G., Isolation of strains degrading poly(vinyl alcohol) at high temperatures and their biodegradation ability. Polymer Degradation and Stability **95**(1), 89-93 **(2010)**

References

79. Shuichi M. and Kazunobu T., in *Hydrogels and Biodegradable Polymers for Bioapplications*, Raphael M. O., Samuel J. H., and Kinam P., Eds, American Chemical Society: Washington, DC **1996**.
80. Chiellini E., Corti A., D'Antone S., and Solar R., Biodegradation of poly(vinyl alcohol) based materials. Progress in Polymer Science **28**(6), 963-1014 **(2003)**
81. Wu Q., Chen N., Li L., and Wang Q., Structure evolution of melt-spun poly(vinyl alcohol) fibers during hot-drawing. Journal of Applied Polymer Science **124**(1), 421-428 **(2012)**
82. Tubbs R. K., Sequence distribution of partially hydrolyzed poly(vinyl acetate). Journal of Polymer Science Part A-1: Polymer Chemistry **4**(3), 623-629 **(1966)**
83. *Airvol Poly(vinyl alcohol) Manual*. 1990, Air Product and Chemicals, Inc.,: Allentown, PA.
84. Finch C.A., *Polyvinyl Alcohol-Developments*, John Wiley & Sons, Inc.: Chichester, New York, Brisbane, Toronto, Singapore **1992**.
85. Fink J. K., in *Handbook of Engineering and Specialty Thermoplastics, Water Soluble Polymers*, Fink J. K., Editor, John Wiley & Sons, Inc. **2011**.
86. Nord F. F., Bier M., and Timasheff Serge N., Investigations on Proteins and Polymers. IV.1 Critical Phenomena in Polyvinyl Alcohol-Acetate Copolymer Solutions. Journal of the American Chemical Society **73**(1), 289-293 **(1951)**
87. Drumright R. E., Gruber P. R., and Henton D. E., Polylactic Acid Technology. Advanced Materials **12**(23), 1841-1846 **(2000)**
88. Hartmann M. H., in *Biopolymers from Renewable Resources*, Kaplan D. L., Editor, Springer Berlin Heidelberg **1998**.
89. Nieuwenhuis J., Synthesis of polylactides, polyglycolides and their copolymers. Clinical Materials **10**(1–2), 59-67 **(1992)**
90. Garlotta D., A Literature Review of Poly(Lactic Acid). Journal of Polymers and the Environment **9**(2), 63-84 **(2001)**
91. Gupta A. P. and Kumar V., New emerging trends in synthetic biodegradable polymers – Polylactide: A critique. European Polymer Journal **43**(10), 4053-4074 **(2007)**
92. Benninga H., *A History of Lactic Acid Making: A Chapter in the History of Biotechnology*, Kluwer Academic **1990**.
93. Carothers W. H., Dorough G. L., and van Natta F. J., Studies of polymerization and ring formation. x. The reversible polymerization of six-membered cyclic esters. Journal of the American Chemical Society **54**(2), 761-772 **(1932)**
94. Lowe C.E., *Preparation of high molecular weight polyhydroxyacetic ester*, Patent No. US2668162 A, Du Pont, US. **1954**.
95. Jiménez A., Peltzer M., and Ruseckaite R., *Poly(lactic acid) Science and Technology: Processing, Properties, Additives and Applications*, Royal Society of Chemistry **2014**.
96. Ren J., *Biodegradable Poly(Lactic Acid): Synthesis, Modification, Processing and Applications*, Springer Berlin Heidelberg **2010**.
97. Lim L.-T., Cink K., and Vanyo T., in *Processing of Poly(Lactic Acid): Synthesis structures properties processing and applications*, Auras R., et al., Eds, John Wiley & Sons, Inc. **2010**.
98. Yang F., et al., Fabrication of nano-structured porous PLLA scaffold intended for nerve tissue engineering. Biomaterials **25**(10), 1891-1900 **(2004)**
99. Yang F., Murugan R., Wang S., and Ramakrishna S., Electrospinning of nano/micro scale poly(l-lactic acid) aligned fibers and their potential in neural tissue engineering. Biomaterials **26**(15), 2603-2610 **(2005)**
100. Sui G., et al., Poly-L-lactic acid/hydroxyapatite hybrid membrane for bone tissue regeneration. Journal of Biomedical Materials Research Part A **82A**(2), 445-454 **(2007)**
101. Zong X., et al., Electrospun fine-textured scaffolds for heart tissue constructs. Biomaterials **26**(26), 5330-5338 **(2005)**

References

102. Thorvaldsson A., Stenhamre H., Gatenholm P., and Walkenström P., Electrospinning of Highly Porous Scaffolds for Cartilage Regeneration. Biomacromolecules **9**(3), 1044-1049 **(2008)**

103. Lee S. J., Yoo J. J., Lim G. J., Atala A., and Stitzel J., In vitro evaluation of electrospun nanofiber scaffolds for vascular graft application. Journal of Biomedical Materials Research Part A **83A**(4), 999-1008 **(2007)**

104. Stitzel J. D., Pawlowski K. J., Wnek G. E., Simpson D. G., and Bowlin G. L., Arterial Smooth Muscle Cell Proliferation on a Novel Biomimicking, Biodegradable Vascular Graft Scaffold. Journal of Biomaterials Applications **16**(1), 22-33 **(2001)**

105. Xu C. Y., Inai R., Kotaki M., and Ramakrishna S., Aligned biodegradable nanofibrous structure: a potential scaffold for blood vessel engineering. Biomaterials **25**(5), 877-886 **(2004)**

106. Inoguchi H., et al., Mechanical responses of a compliant electrospun poly(l-lactide-co-ε-caprolactone) small-diameter vascular graft. Biomaterials **27**(8), 1470-1478 **(2006)**

107. Vaz C. M., van Tuijl S., Bouten C. V. C., and Baaijens F. P. T., Design of scaffolds for blood vessel tissue engineering using a multi-layering electrospinning technique. Acta Biomaterialia **1**(5), 575-582 **(2005)**

108. In Jeong S., et al., Tissue-engineered vascular grafts composed of marine collagen and PLGA fibers using pulsatile perfusion bioreactors. Biomaterials **28**(6), 1115-1122 **(2007)**

109. Telemeco T. A., et al., Regulation of cellular infiltration into tissue engineering scaffolds composed of submicron diameter fibrils produced by electrospinning. Acta Biomaterialia **1**(4), 377-385 **(2005)**

110. He Y., Zhu B., and Inoue Y., Hydrogen bonds in polymer blends. Progress in Polymer Science **29**(10), 1021-1051 **(2004)**

111. Paul D. R., in *Polymer Blends and Mixtures*, Walsh D. J., Higgins J. S., and Maconnachie A., Eds, Springer Netherlands **1985**.

112. Utracki L. A., Walsh D. J., and Weiss R. A., in *Multiphase Polymers: Blends and Ionomers*, Utracki L. A. and Weiss R. A., Eds, American Chemical Society **1989**.

113. Macosko C. W., Morphology development and control in immiscible polymer blends. Macromolecular Symposia **149**(1), 171-184 **(2000)**

114. Thomas S., Groeninckx G., and Harrats C., in *Micro- and Nanostructured Multiphase Polymer Blend Systems: Phase Morphology and Interfaces*, Harrats C., Thomas S., and Groeninckx G., Eds, CRC Press: Boca Raton, London, New York **2006**.

115. Han C.D., in *Multiphase Flow in Polymer Processing*, C.D. Han, Editor, Academic Press: New York, London, Toronto, Sydney, San Francisco **1981**.

116. Fakirov S., Duhovic M., Maitrot P., and Bhattacharyya D., From PET Nanofibrils to Nanofibrillar Single-Polymer Composites. Macromolecular Materials and Engineering **295**(6), 515-518 **(2010)**

117. Duhovic M., Fakirov S., Holschuh R., Mitschang P., and Bhattacharyya D., in *Synthetic Polymer-Polymer Composites*, Bhattacharyya D. and Fakirov S., Eds, Carl Hanser Verlag GmbH & Co. KG: Munich, Germany **2012**.

118. Kim J. Y., Kim S. H., and Kikutani T., Fiber property and structure development of polyester blend fibers reinforced with a thermotropic liquid-crystal polymer. Journal of Polymer Science Part B: Polymer Physics **42**(3), 395-403 **(2004)**

119. Kim S. Y., Kim S. H., Lee S. H., and Youn J. R., Internal structure and physical properties of thermotropic liquid crystal polymer/poly(ethylene 2,6-naphthalate) composite fibers. Composites Part A: Applied Science and Manufacturing **40**(5), 607-612 **(2009)**

120. McCardle R., Bhattacharyya D., and Fakirov S., Effect of Reinforcement Orientation on the Mechanical Properties of Microfibrillar PP/PET and PET Single-Polymer Composites. Macromolecular Materials and Engineering **297**(7), 711-723 **(2012)**

121. Sombatdee S., Amornsakchai T., and Saikrasun S., Reinforcing performance of recycled PET microfibrils in comparison with liquid crystalline polymer for polypropylene based composite fibers. Journal of Polymer Research **19**(3), 1-13 **(2012)**

References

122. Tjong S. C., Structure, morphology, mechanical and thermal characteristics of the in situ composites based on liquid crystalline polymers and thermoplastics. Materials Science and Engineering: R: Reports **41**(1–2), 1-60 **(2003)**

123. Saikrasun S., Limpisawasdi P., and Amornsakchai T., Comparative study on phase and properties between rPET/PS and LCP/PS in situ microfibrillar-reinforced composites. Journal of Polymer Research **16**(4), 443-454 **(2009)**

124. Varma D. S. and Dhar V. K., Nylon 6/PET Polymer Blends: Mechanical Properties of Fibers. Textile Research Journal **58**(5), 274-279 **(1988)**

125. Xing Q., et al., In situ gradient nano-scale fibril formation during polypropylene (PP)/polystyrene (PS) composite fine fiber processing. Polymer **46**(14), 5406-5416 **(2005)**

126. Favis B. D., in *Polymer Blends: Formulation and Performance, Performance*, Paul D.R. and Bucknall C.B., Eds, Wiley: New York, Chichester, Weinheim, Brisbane, Singapore, Toronto **2000**.

127. Han C.D. and Kim Y. W., Studies on melt spinning. V. Elongational viscosity and spinnability of two-phase systems. Journal of Applied Polymer Science **18**(9), 2589-2603 **(1974)**

128. Petrie C.J.S., *Elongational flows: aspects of the behaviour of model elasticoviscous fluids*, Pitman Publishing **1979**.

129. Gupta V.B. and Bhuvanesh Y.C., in *Manufactured Fibre Technology*, Gupta V.B. and Kothari V.K., Eds, Springer Netherlands **1997**.

130. Lin C.-A. and Ku T.-H., Shear and elongational flow properties of thermoplastic polyvinyl alcohol melts with different plasticizer contents and degrees of polymerization. Journal of Materials Processing Technology **200**(1–3), 331-338 **(2008)**

131. Taylor G. I., The Viscosity of a Fluid Containing Small Drops of Another Fluid. Proc. R. Soc. **138**(834), 41-48 **(1932)**

132. Taylor G. I., The Formation of Emulsions in Definable Fields of Flow. Proc. R. Soc. **146**(858), 501-523 **(1934)**

133. Cotto D., Saillard P., Agassant J.F., and Haudin J.M., in *Interrelations between Processing, Structure and Properties of Polymeric Materials*, Siferis J.C and Theocaris P.S., Eds, Elsevier Amsterdam **1984**.

134. Fortelný I., in *Micro- and Nanostructured Multiphase Polymer Blend Systems: Phase Morphology and Interfaces*, Harrats C.., Thomas S., and Groeninckx G., Eds, CRC Press: Boca Ranton, London, New York **2006**.

135. Cox R. G., The deformation of a drop in a general time-dependent fluid flow. Journal of Fluid Mechanics **37**(03), 601-623 **(1969)**

136. Utracki L. A. and Shi Z. H., Development of polymer blend morphology during compounding in a twin-screw extruder. Part I: Droplet dispersion and coalescence-a review. Polymer Engineering & Science **32**(24), 1824-1833 **(1992)**

137. De Bruijn R. A., *Deformation and breakup of drops in simple shear flows*, Eindhoven University of Technology, Ph.D. Thesis: The Netherlands **1989**.

138. Sundararaj U., in *Micro- and Nanostructured Multiphase Polymer Blend Systems: Phase Morphology and Interfaces*, Harrats C., Thomas S., and Groeninckx G., Eds, Taylor & Francis: Boca Raton, London, New York **2006**.

139. Grace† H. P., Dispersion phenomena in high viscosity immiscible fluid systems and application of static mixers as dispersion devices in such systems. Chemical Engineering Communications **14**(3-6), 225-277 **(1982)**

140. Carothers W. H. and Hill J. W., Studies of polymerization and ring formation. xv. Artificial fibers from synthetic linear condensation superpolymers Journal of the American Chemical Society **54**(4), 1579-1587 **(1932)**

141. Carothers W. H. and van Natta F. J., Studies of Polymerization and Ring Formation. xviii Polyesters from ι-Hydroxydecanoic Acid. Journal of the American Chemical Society **55**(11), 4714-4719 **(1933)**

References

142. Ziabicki A. and Kedzierska K., Studies on the orientation phenomena by fiber formation from polymer melts. Part I. Preliminary investigations on polycapronamide. Journal of Applied Polymer Science 2(4), 14-23 **(1959)**

143. Ziabicki A., Studies on orientation phenomena by fiber formation from polymer melts. Part II. Theoretical considerations. Journal of Applied Polymer Science 2(4), 24-31 **(1959)**

144. Ziabicki A. and Kedzierska K., Mechanical aspects of fibre spinning process in molten polymers. Part I. Stream Diameter and Velocity Distribution along the Spinning Way. Kolloid-Zeitschrift 171(2), 111-119 **(1960)**

145. Ziabicki A. and Kedzierska K., Mechanical aspects of fibre spinning process in molten polymers. Part II. Stream Broadening after the Exit from the Channel of Spinneret. Kolloid-Zeitschrift 171(1), 51-61 **(1960)**

146. Ziabicki A., Mechanical aspects of fibre spinning process in molten polymers. Part III. Tensile Force and Stress. Kolloid-Zeitschrift 175(1), 14-27 **(1961)**

147. Ziabicki A., Differential equations for velocity components in fibre spinning process. Kolloid-Zeitschrift 179(2), 116-117 **(1961)**

148. Ziabicki A. and Kedzierska K., Studies on the orientation phenomena by fiber formation from polymer melts. III. Effect of structure on orientation. Condensation polymers. Journal of Applied Polymer Science 6(19), 111-119 **(1962)**

149. Ziabicki A. and Kedzierska K., Studies on the orientation phenomena by fiber formation from polymer melts. IV. Effect of molecular structure on orientation. Polyethylene and polystyrene. Journal of Applied Polymer Science 6(21), 361-367 **(1962)**

150. Ziabicki A. and Takserman-Krozer R., Effect of rheological factors on the length of liquid threads. Kolloid-Zeitschrift und Zeitschrift für Polymere 199(1), 9-13 **(1964)**

151. Ziabicki A., Jarecki L., and Wasiak A., Dynamic modelling of melt spinning. Computational and Theoretical Polymer Science 8(1–2), 143-157 **(1998)**

152. Ziabicki A. and Takserman-Krozer R., Mechanism of breakage of liquid threads. Kolloid-Zeitschrift und Zeitschrift für Polymere 198(1-2), 60-65 **(1964)**

153. Kase S. and Matsuo T., Studies on melt spinning. I. Fundamental equations on the dynamics of melt spinning. Journal of Polymer Science Part A: General Papers 3(7), 2541-2554 **(1965)**

154. Kase S. and Matsuo T., Studies on melt spinning. II. Steady-state and transient solutions of fundamental equations compared with experimental results. Journal of Applied Polymer Science 11(2), 251-287 **(1967)**

155. Kase S., Studies on melt spinning. III. Velocity field within the thread. Journal of Applied Polymer Science 18(11), 3267-3278 **(1974)**

156. Kase S., Studies on melt spinning. IV. On the stability of melt spinning. Journal of Applied Polymer Science 18(11), 3279-3304 **(1974)**

157. Matsuo T. and Kase S., Studies on melt spinning. VII. Temperature profile within the filament. Journal of Applied Polymer Science 20(2), 367-376 **(1976)**

158. Kase S. and Araki M., Studies on melt spinning. VIII. Transfer function approach. Journal of Applied Polymer Science 27(11), 4439-4465 **(1982)**

159. Han C. D., A theoretical study on fiber spinnability. Rheologica Acta 9(3), 355-365 **(1970)**

160. Han C.D. and Lamonte R. R., Studies on Melt Spinning. I. Effect of Molecular Structure and Molecular Weight Distribution on Elongational Viscosity. Transactions of The Society of Rheology 16(3), 447-472 **(1972)**

161. Lamonte R. R. and Han C.D., Studies on melt spinning. II. Analysis of the deformation and heat transfer processes. Journal of Applied Polymer Science 16(12), 3285-3306 **(1972)**

162. Han C.D., Lamonte R. R., and Shah Y. T., Studies on melt spinning. III. Flow instabilities in melt spinning: Melt fracture and draw resonance. Journal of Applied Polymer Science 16(12), 3307-3323 **(1972)**

163. Han C.D., Lamonte R. R., and Drexler L. H., Studies on melt spinning. IV. Spinning through a ribbon die. Journal of Applied Polymer Science 17(4), 1165-1172 **(1973)**

References

164. Han C. D. and Kim Y.W., Studies on melt spinning. VI. The effect of deformation history on elongational viscosity, spinnability, and thread instability. Journal of Applied Polymer Science **20**(6), 1555-1571 **(1976)**

165. Kim Y. W. and Han C. D., Studies on melt spinning. VII. Elongational viscosity and fiber morphology of multiphase polymer systems. Journal of Applied Polymer Science **21**(2), 515-524 **(1977)**

166. Han C. D. and Apte S. M., Studies on melt spinning. VIII. The effects of molecular structure and cooling conditions on the severity of draw resonance. Journal of Applied Polymer Science **24**(1), 61-87 **(1979)**

167. Andrews E. H., Cooling of a spinning thread-line. British Journal of Applied Physics **10**(2), 104 **(1959)**

168. Hamana I., *Der Verlauf der Fadenbildung beim Schmelzspinnen*. 1968, Lenzing Berichte. p. 118-132.

169. George H. H., Model of steady-state melt spinning at intermediate take-up speeds. Polymer Engineering & Science **22**(5), 292-299 **(1982)**

170. Lin L. C. T. and Hauenstein J., Cooling and attenuation of threadline in melt spinning of poly(ethylene terephthalate). Journal of Applied Polymer Science **18**(12), 3509-3521 **(1974)**

171. Denn M. M., in *Computational Analysis of Polymer Processing*, Pearson J. R. A. and Richardson S. M., Eds, Springer Netherlands **1983**.

172. Ziabicki A., *Fundamentals of fibre formation: the science of fibre spinning and drawing*, John Wiley & Sons Ltd **1976**.

173. Ziabicki A. and Kawai H., *High-speed fiber spinning: science and engineering aspects*, Wiley & Sons: New York, Chichester, Brisbane, Toronto, Singapore **1985**.

174. Nakajima T., Kajiwara K., and McIntyre J.E., *Advanced Fiber Spinning Technology*, Taylor & Francis: Oxford, Cambridge, New Delhi, **1994**.

175. Beyreuther R. and Brünig H., *Dynamics of Fibre Formation and Processing: Modelling and Application in Fibre and Textile Industry*, Springer Berlin Heidelberg **2007**.

176. Golzar M., *Melt Spinning of the Fine PEEK Filaments*, Dr.-Ing. Thesis,Technischen Universität Dresden, Germany **2004**.

177. Glicksman L. R., *An investigation of the shape, temperature distribution and tension of a heated free jet flowing at ultra low Reynolds numbers*, Massachusetts Institute of Technology, Ph.D. Thesis **1964**.

178. Glicksman L. R., The Cooling of Glass Fibres. Glass Technology **9**(5), 131-138 **(1968)**

179. Ohkoshi Y., Konda A., Kikutani T., and Shimizu J., Melt Spinning of Poly Ether-ether-ketone (PEEK). Cooling, Thinning, and Structure Development on Spin-line. Sen'i Gakkaishi **49**(5), 211-219 **(1993)**

180. Ohkoshi Y., et al., Cooling Behavior of the Spinning Line of Poly(ether ether ketone). Sen'i Gakkaishi **56**(7), 340-347 **(2000)**

181. Brünig H., Beyreuther R., and Hoffman H., The influence of quench air on fiber formation and properties in the melt spinning process. Intern. Fiber J. **14**(4), 104-107 **(1999)**

182. Kuraray Co. Ltd. *Thermoplastic processable polyvinyl alcohol*. Available from: http://www.kuraray.eu/fileadmin/Downloads/Mowiflex/Mowiflex_TC_flyer.pdf.

183. Dorigato A. and Pegoretti A., Biodegradable single-polymer composites from polyvinyl alcohol. Colloid and Polymer Science **290**(4), 359-370 **(2012)**

184. Kuraray Co. Ltd. *Technical data sheet*. Available from: http://www.kuraray.eu/fileadmin/Downloads/Mowiflex/technical_data_sheets/TDS_Mowiflex_en.pdf.

185. Carreau P.J., De Kee D., and Chhabra R.P., in *Rheology of polymeric systems*, Carreau P.J., De Kee D., and Chhabra R.P., Eds, Hanser: Munich, Vienna, New York **1997**.

186. Randall J., *NatureWorks launches new Ingeo grades for improved fiber properties*. 2014, NatureWorks LLC.

187. *Unstruction Manual: Model LS50/LS50PT Multiplexed LaserSpeed*, T. Incorporated, Editor. 1995: USA.

188. Durst F., Melling A., and Whitelaw J.H., *Principles and practice of laser-Doppler anemometry*, Academic Press **1976**.

References

189. Albrecht H.E., *Laser Doppler and Phase Doppler Measurement Techniques*, Springer **2003**.

190. Golzar M., Beyreuther R., Brünig H., Tändler B., and Vogel R., Online temperature measurement and simultaneous diameter estimation of fibers by thermography of the spinline in the melt spinning process. Advances in Polymer Technology 23(3), 176-185 **(2004)**

191. Vogel R., Hatzikiriakos S. G., Brünig H., Tändler B., and Golzar M., Improved Spinnability of Metallocene Polyethylenes by Using Processing Aids. International Polymer Processing 18(1), 67-73 **(2003)**

192. Ishibashi T., Aoki K., and Ishii T., Studies on melt spinning of nylon 6. I. Cooling and deformation behavior and orientation of nylon 6 threadline. Journal of Applied Polymer Science 14(6), 1597-1613 **(1970)**

193. Oh T., Studies on melt spinning process of hollow polyethylene terephthalate fibers. Polymer Engineering & Science 46(5), 609-616 **(2006)**

194. González-Núnez R., De Kee D., and Favis B. D., The influence of coalescence on the morphology of the minor phase in melt-drawn polyamide-6/HDPE blends. Polymer 37(21), 4689-4693 **(1996)**

195. Merkus H. K., in *Particle Size Measurements*, Merkus H. K., Editor, Springer Netherlands **2009**.

196. Blott S. J. and Pye K., Particle shape: a review and new methods of characterization and classification. Sedimentology 55(1), 31-63 **(2008)**

197. Karger-Kocsis J., Kalló A., and Kuleznev V. N., Phase structure of impact-modified polypropylene blends. Polymer 25(2), 279-286 **(1984)**

198. Fakirov S., Nano-/microfibrillar polymer–polymer and single polymer composites: The converting instead of adding concept. Composites Science and Technology 89(0), 211-225 **(2013)**

199. Emilio M., Natalia H.-M., Ester Z., and Jose-Ramon S., in *Characterization of Polymer Blends: Miscibility, Morphology and Interfaces*, Thomas S., Grohens Y., and Jyotishkumar P., Eds, Wiley **2014**.

200. Tsuji H. and Muramatsu H., Blends of aliphatic polyesters: V. non-enzymatic and enzymatic hydrolysis of blends from hydrophobic poly(l-lactide) and hydrophilic poly(vinyl alcohol). Polymer Degradation and Stability 71(3), 403-413 **(2001)**

201. Jawalkar S. S. and Aminabhavi T. M., Molecular modeling simulations and thermodynamic approaches to investigate compatibility/incompatibility of poly(l-lactide) and poly(vinyl alcohol) blends. Polymer 47(23), 8061-8071 **(2006)**

202. Kuraray America Inc., *Thermoplastic Processing of Mowiol / PVOH*. Technical Data Sheet.

203. Mansur H. S., Oréfice R. L., and Mansur A. A. P., Characterization of poly(vinyl alcohol)/poly(ethylene glycol) hydrogels and PVA-derived hybrids by small-angle X-ray scattering and FTIR spectroscopy. Polymer 45(21), 7193-7202 **(2004)**

204. Andrade G., Barbosa-Stancioli E. F., Mansur A. A. P., Vasconcelos W. L., and Mansur H. S., Design of novel hybrid organic–inorganic nanostructured biomaterials for immunoassay applications. Biomedical Materials 1(4), 221 **(2006)**

205. Mansur H. S., Sadahira C. M., Souza A. N., and Mansur A. A. P., FTIR spectroscopy characterization of poly(vinyl alcohol) hydrogel with different hydrolysis degree and chemically crosslinked with glutaraldehyde. Materials Science and Engineering: C 28(4), 539-548 **(2008)**

206. Mallapragada S. K. and Peppas N. A., Dissolution mechanism of semicrystalline poly(vinyl alcohol) in water. Journal of Polymer Science Part B: Polymer Physics 34(7), 1339-1346 **(1996)**

207. Costa H. S., et al., Sol–gel derived composite from bioactive glass–polyvinyl alcohol. Journal of Materials Science 43(2), 494-502 **(2008)**

208. Auras R., Harte B., and Selke S., An Overview of Polylactides as Packaging Materials. Macromolecular Bioscience 4(9), 835-864 **(2004)**

209. Kister G., Cassanas G., and Vert M., Effects of morphology, conformation and configuration on the IR and Raman spectra of various poly(lactic acid)s. Polymer 39(2), 267-273 **(1998)**

210. Gonçalves C. M. B., Coutinho J. A. P., and Marrucho I. M., in *Poly(Lactic Acid): Synthesis, Structures, Properties, Processing, and Applications*, Auras R., et al., Eds, John Wiley & Sons **2010**.

References

211. Nibbering E. J., et al., in *Analysis and Control of Ultrafast Photoinduced Reactions*, Kühn O. and Wöste L., Eds, Springer Berlin Heidelberg **2007**.

212. Radjabian M., Kish M. H., and Mohammadi N., Characterization of poly(lactic acid) multifilament yarns. I. The structure and thermal behavior. Journal of Applied Polymer Science 117(3), 1516-1525 **(2010)**

213. Hikasa J.-i., in *Kirk-Othmer Encyclopedia of Chemical Technology*, Herman F. M., Editor, John Wiley & Sons, Inc. **2000**.

214. Sakurada I., in *Handbook of fiber science and technology. vol. IV. fiber chemistry*, Lewin M. and Pearce E. M., Eds, Marcel Dekker: New York **1985**.

215. Izard E.F., *Process of forming films, threads, and the like*, Patent No. US2169250A, Du Pont, US. **1939**.

216. Izard E.F. and Kohn J., *Method of making films, threads, and the like*, Patent No. US2236061A, Du Pont, US. **1941**.

217. Marten F. L., in *Encyclopedia of Polymer Science and Technology*, Mark H.F. and Kroschwitz J.I., Eds, John Wiley & Sons **2003**.

218. Kwon Y.D., Kavesh S., and Prevorsek D.C., *High strength and modulus polyvinyl alcohol fibers and method of their preparation*, Patent No. 4603083, Allied Corporation. **1986**.

219. Nishino T., K. S., Gotoh K., and Nakamae K., Melt processing of poly(vinyl alcohol) through blending with sugar pendant polymer. Polymer 43(9), 2869-2873 **(2002)**

220. Carreau P. J., Rheological Equations from Molecular Network Theories. Transactions of The Society of Rheology (1957-1977) 16(1), 99-127 **(1972)**

221. Yasuda K., *Investigation of the Analogies Between Viscometric and Linear Viscoelastic Properties of Polystyrene Fluids*, Massachusetts Institute of Technology, Department of Chemical Engineering. Ph.D. Thesis, US **1979**.

222. Meredith R., 10—The tensile behaviour of raw cotton and other textile fibres. Journal of the Textile Institute Transactions 36(5), T107-T130 **(1945)**

223. Morton W. E. and Hearle J. W. S., in *Physical Properties of Textile Fibres*, Morton W. E. and Hearle J. W. S., Eds, Woodhead Publishing **2008**.

224. Lin S. T., Bhattacharyya D., Fakirov S., Matthews B., and Cornish J., A Novel Microfibrillar Composite Approach towards Manufacturing Nanoporous Tissue Scaffolds. Mechanics of Advanced Materials and Structures 21(3), 237-243 **(2013)**

225. Pearson J. R. A. and Matovich M. A., Spinning a Molten Threadline. Steady-State Isothermal Viscous Flows. Industrial & Engineering Chemistry Fundamentals 8(3), 512-520 **(1969)**

226. Shah Y. T. and Pearson J. R. A., On the Stability of Nonisothermal Fiber Spinning. Industrial & Engineering Chemistry Fundamentals 11(2), 145-149 **(1972)**

227. Papanastasiou T. C., Macosko C. W., Scriven L. E., and Chen Z., Fiber spinning of viscoelastic liquid. AIChE Journal 33(5), 834-842 **(1987)**

228. Ishizuka O. and Koyama K., in *High-speed fiber spinning: science and engineering aspects*, , Ziabicki A. and Kawai H., Eds, Wiley & Sons: New York, Chichester, Brisbane, Toronto, **1985**.

229. Gianelli W., Camino G., Dintcheva N. T., V.S. L, and Mantia F. P. L., EVA-Montmorillonite Nanocomposites: Effect of Processing Conditions. Macromolecular Materials and Engineering 289(3), 238-244 **(2004)**

230. Bruno V. and Wiboon L., in *Polymer Nanocomposite Research Advances*, Thomas S. and Zaikov G. E., Eds, Nova Science Pub Inc **2008**.

231. Keller A., Unusual orientation phenomena in polyethylene interpreted in terms of the morphology. Journal of Polymer Science 15(79), 31-49 **(1955)**

232. Spruiell J. E. and J.L. White, Structure development during polymer processing: Studies of the melt spinning of polyethylene and polypropylene fibers. Polymer Engineering & Science 15(9), 660-667 **(1975)**

233. Harrats C., in *Multiphase Polymer-Based Materials: An Atlas of Phase Morphology at the Nano and Micro Scale*, Harrats C., Editor, CRC Press: Boca Raton, London, New York, **2009**.

References

234. Favis B. D. and Chalifoux J. P., Influence of composition on the morphology of polypropylene /polycarbonate blends. Polymer **29**(10), 1761-1767 **(1988)**

235. Bourry D. and Favis B. D., Morphology development in a polyethylene/polystyrene binary blend during twin-screw extrusion. Polymer **39**(10), 1851-1856 **(1998)**

236. Heindl M., Sommer M.-K., and Münstedt H., Morphology development in polystyrene/polyethylene blends during uniaxial elongational flow. Rheologica Acta **44**(1), 55-70 **(2004)**

237. Wu S., Formation of dispersed phase in incompatible polymer blends: Interfacial and rheological effects. Polymer Engineering & Science **27**(5), 335-343 **(1987)**

238. Suparno M., Dolan K. D., Ng P. K. W., and Steffe J. F., Average shear rate in a twin-screw extruder as a function of degree of fill, flow behavior index, screw speed and screw configuration. Journal of Food Process Engineering **34**(4), 961-982 **(2011)**

239. Burkhardt K., Herrmann H., and Jakopin S., Plasticating and Mixing Principles of Intermeshing Corotatingand Counter-rotating Twin Screw Extruders. SPE ANTEC Tech. Papers **36**498 **(1978)**

240. Sundararaj U. and Macosko C. W., Drop Breakup and Coalescence in Polymer Blends: The Effects of Concentration and Compatibilization. Macromolecules **28**(8), 2647-2657 **(1995)**

241. Roland C. M. and Böuhm G. G. A., Shear-induced coalescence in two-phase polymeric systems. I. Determination from small-angle neutron scattering measurements. Journal of Polymer Science: Polymer Physics Edition **22**(1), 79-93 **(1984)**

242. Plochocki A. P., Dagli S. S., and Andrews R. D., The interface in binary mixtures of polymers containing a corresponding block copolymer: Effects of industrial mixing processes and of coalescence. Polymer Engineering & Science **30**(12), 741-752 **(1990)**

243. Favis B. D. and Chalifoux J. P., The effect of viscosity ratio on the morphology of polypropylene /polycarbonate blends during processing. Polymer Engineering & Science **27**(21), 1591-1600 **(1987)**

244. Allan R. S. and Mason S. G., Particle motions in sheared suspensions. XIV. Coalescence of liquid drops in electric and shear fields. Journal of Colloid Science **17**(4), 383-408 **(1962)**

List of figures

Figure 1.1 Schematic representation of cross- and longitudinal section of blend fibers/MFFs (a, b), and micro-/nanofibrils (c).. 2

Figure 2.1 Chemical representations of PVAL 8

Figure 2.2 Solubility of a PVAL sample with the degree of polymerization of 1700 in the relation with the degree of hydrolysis at various dissolution temperatures (adapted from Ref. [69], with permission from John Wiley and Sons)... 10

Figure 2.3 Solubility of various PVAL in water vs. solution temperature (adapted from Ref. [69], with permission from John Wiley and Sons)... 10

Figure 2.4 Reaction routes of producing PLA from lactic acid [63, 87, 95, 96] 11

Figure 2.5 Different types of useful morphologies of immiscible polymer blends (reproduced from Ref. [113], with permission from John Wiley and Sons) 13

Figure 2.6 Droplet deformation in simple shear (a) and in plane hyperbolic flow fields (b), modified after [132]... 16

Figure 2.7 Critical capillary number Ca_c vs. viscosity λ ratio for droplet breakup in shear and hyperbolic flow fields, adapted from Ref. [139] 19

Figure 2.8 A schematic diagram of fiber formation in melt spinning process 20

Figure 2.9 A schematic draw of forces acting on a filament, adapted from Ref. [175] 22

Figure 3.1 Heat flow curves of PVAL.. 27

Figure 3.2 Heat flow curves of PLA.. 29

Figure 3.3 Schematic drawing of piston type extrusion spinning equipment................................ 31

Figure 3.4 Principle of melt spinning device in IPF Dresden [175]... 32

Figure 3.5 Scheme of the drawing process (above) and drawing device at IPF Dresden (below) 33

Figure 3.6 Model LS50M multiplexed LaserSpeed® system 35

List of figures

Figure 3.7 Layout of optical path to the filament, modified after [187] 36

Figure 3.8 A histograph of velocity values with 100 data points using LDV technique 36

Figure 3.9 Photographic of test stand for temperature measurement using infrared camera 37

Figure 3.10 Photographic of test stand for measurement the emissivity as correction factor 39

Figure 3.11 The emissivity vs. the diameter of PLA/PVA blends in comparison with pure PEEK filament (*) adapted from Ref. [190] 39

Figure 3.12 A snapshot of filament temperature measurement using the infrared camera (a) and the three temperature profiles perpendicular to spinline (b) 40

Figure 3.13 Fiber-capturing device and schematic view of a captured molten filament 41

Figure 3.14 Measurement of the filament diameter in the cross-section of PLA/PVAL filaments using microscopy (left) and the photograph of filament keeping device (right) 42

Figure 3.15 Measurement of the filament diameter in the longitudinal direction (above) and the photograph of a PEAK glass and the two captured filaments obtained using fiber capturing device (below) 42

Figure 3.16 Preparation of PLA scaffolds for the tensile testing: (a) self-constructed frame, (b) off-line drawn PLA/PVAL filaments wound and fixed in the frame, (c) the filaments are then immersed in water during removing process 43

Figure 3.17 Thickness measurement of a PLA scaffold 43

Figure 3.18 PVAL removing process in distiller water: samples were fixed in filament-keeping device (a and b), then were immersed in water for 24 hours. 44

Figure 3.19 An original SEM image of PLA/PVAL blend (a) and the image analyzed using SIA software (b): The blue domains and grey background represents the dispersed PLA phase and PVAL matrix, respectively. 45

Figure 3.20 Illustrating of circular equivalent diameter (left) and circularity of a 2D particles (right) 46

Figure 3.21 An example of the measurement of dispersed PLA phase determined using SIA 46

Figure 4.1 DSC thermograms obtained from the 2nd heating runs of neat PLA, PVAL, and PLA/PVAL blends: (a) heat flow vs. temperature, (b) derivative heat flow vs. temperature in the glass transition range 49

Figure 4.2 The dependence of glass transition temperature Tg on the blend ratios 50

Figure 4.3 Chemical structure of poly (vinyl alcohol-*co*-vinyl acetate) 50

Figure 4.4 ATR-FTIR spectra of PVAL (a) and PLA (b) 51

List of figures

Figure 4.5 Various potential hydroxyl- hydroxyl (1) and hydroxyl-carbonyl (2) hydrogen-bonds between PVAL and PLA 52

Figure 4.6 ATR-FTIR spectra comparison of PVAL/PLA blends with PVAL (a) and with PLA (b) over the range of 4000 − 1500 cm^{-1} 53

Figure 4.7 The ATR-FTIR spectra (top) and the second derivatives of the ATR-FTIR spectra (bottom) of the PVAL, PLA and their blends in the carbonyl region: (a) PVAL, PLA/PVAL blends, (b) PLA, PLA/PVAL blends 54

Figure 4.8 SEM images of fracture surfaces of the as-extruded PLA/PVAL blend after etching of the PLA phase at various PLA/PVAL blend ratios: scale bars for the left and right column are 10 µm and 1 µm, respectively 55

Figure 4.9 SEM mages of a microtome surface of PLA/PVAL 30/70 blend without etching the PLA phase: (a) scale bar: 10 µm, (b) scale bar: 1 µm 56

Figure 4.10 AMF images of a microtome surface of PLA/PVAL 30/70 blend: (a) scale bar: 4 µm; (b) scale bar: 800 nm 56

Figure 4.11 Experimental setup on microcompounder (adapted from the oral presentation at the 30th international conference of the Polymer Processing Society, 08-12 June, 2014, Cleveland, Ohio, USA) 58

Figure 4.12 SEM images of PLA structures prepared from the as-extruded PLA/PVAL blends that were obtained using the microcompounder (left column, scale bar: 10 µm, except Figure 4.12a with scale bar 1 µm), and from the PLA/PVAL blend filaments (right column, scale bar: 1 µm) 59

Figure 4.13 Photographs of blend filaments before (above) and after (below) removing PVAL matrix 59

Figure 4.14 Photographs of molten PLA/PVAL 30/70 in the spinning head after two hours of an unsuccessful trial spinning process 60

Figure 4.15 Mass vs. temperature (a) and derivative mass vs. temperature (b) of PLA, PVAL, and their blends in TG measurement 61

Figure 4.16 Mass vs. time in isothermal TG measurement of PLA, PVAL, and their blends at 200 °C 62

Figure 4.17 Mass vs. time in isothermal TG measurement of PLA/PVAL 30/70 blend at temperatures of 185, 190, 195, 200, and 205 °C 62

Figure 4.18 Complex viscosity vs. frequency for PLA, PVAL, and PLA/PVAL blends (a) and viscosity ratio of PLA/PVAL (b) at 195 °C 63

List of figures

Figure 4.19 Complex viscosity vs. frequency for PLA/PVAL 30/70 blend at different temperatures. 64

Figure 4.20 PLA/PVAL blend melt stability followed in oscillation at different temperatures at constant frequency of 10 rad·s^{-1} 64

Figure 4.21 Tensile and elongation properties of PLA/PVAL filaments for different spinning velocities: Stress-strain curves (a), tenacity and elongation at break (b). 65

Figure 4.22 Tensile and elongation properties of PLA/PVAL filaments for different draw ratios: Stress-strain curves (a), tenacity and elongation at break (b). 66

Figure 4.23 Tenacity and Young's modulus versus draw ratio 67

Figure 4.24 Dependence of the elongation at break on the draw ratio 67

Figure 4.25 Stress-strain curve of PLA/PVAL drawn filaments vs. stress-strain curves of various fibers, modified after [222, 223] 68

Figure 4.26 Photographic image of PLA/PVAL 30/70 filament yarns at different take-up velocity 30, 40, and 50 m·min^{-1} (3 spools on the left) and off-line drawn filament yarns with different draw ratios (5 spools on the right) 69

Figure 4.27 Photographic images of woven (a) and knitted fabrics (b) from PLA/PVAL 70/30 blend filaments 69

Figure 4.28 Photographic image the dried PLA nanofibrillar structures 69

Figure 4.29 SEM images of the remaining PLA nanofibers from PLA/PVAL blend filaments after removing the PVAL matrix: (a) v=30 m·min^{-1}; (b) v=40 m·min^{-1}; (c) v=50 m·min^{-1}; (a1) v=30 m·min^{-1} and DR=1.25; (a2) v=30 m·min^{-1} and DR=1.50; (a3) v=30 m·min^{-1} and DR=1.70; scale bar: 1 μm ... 70

Figure 4.30 Average diameter of PLA fibrils vs. take-up velocity (a) and draw ratio (b) 71

Figure 4.31 PLA fibrils from drawn PLA/PVAL filaments with DR=1.5 (a) and DR=1.7 (b). The fracture of PLA fibrils is marked by a red colour oval. Scale bar: 200 nm 71

Figure 4.32 Frequency distribution histograms of the diameters of PLA fibrils: (a) v=30 m·min^{-1}; (b) v=40 m·min^{-1}; (c) v=50 m·min^{-1}; (a1) v=30 m·min^{-1} and DR=1.25; (a2) v=30 m·min^{-1} and DR=1.50; (a3) v=30 m·min^{-1} 72

Figure 4.33 Cumulative distributions of the diameters of PLA fibrils for various take-up velocities (a) and various draw ratios (b) 72

Figure 4.34 ATR-FTIR spectra of neat PLA, PVAL, PLA/PVAL 30/70 blend, and the PLA scaffolds obtained from PLA/PVAL blend filaments, which were immersed in water bath at room temperature ca. 25 °C for 1, 5, and 10 days. 73

List of figures

Figure 4.35 ATR-FTIR spectra of neat PLA, PVAL, PLA/PVAL 30/70 blend, and the PLA scaffolds obtained from PLA/PVAL blend filaments, which were immersed in water bath at 60 °C for 1, 5, and 10 days. 74

Figure 4.36 Stress-strain curves of 10 different scaffolds from drawn PLA/PVAL filaments............. 75

Figure 4.37 Scaffolds before and after mechanical testing.. 75

Figure 4.38 Filament temperature vs. distance for the take-up velocity of 50 m·min^{-1} and the mass flow rate of 1 g·min^{-1}: (×) uncorrected temperatures obtained using the infrared camera, (●) corrected temperatures, and (— — —) fitted temperature profile using an exponential decay function. 78

Figure 4.39 Filament temperature vs. distance for the take-up velocity of 50 m·min^{-1} and the different mass flow rates: 0.5, 1.0, 1.5, and 2.0 g·min^{-1}. 79

Figure 4.40 Filament temperature vs. distance for the different take-up velocities at constant mass flow rate of 1.0 g·min^{-1}. 79

Figure 4.41 Velocity vs. distance for the specific melt spinning condition: v=50 m·min^{-1}; Q=1 g·min^{-1} 80

Figure 4.42 Velocity (a) and axial strain rate (ASR) (b) vs. distance for the melt spinning conditions: v=50 m·min^{-1}; Q=0.5, 1.0, 1.5, and 2.0 g·min^{-1} 81

Figure 4.43 Velocity (a) and ASR (b) vs. distance for the melt spinning conditions: Q=1.0 g·min^{-1}; v= 10, 30, 50, and 70 m·min^{-1} 82

Figure 4.44 Filament diameter vs. distance for the spinning condition: Q=1.0 g·min^{-1}; v= 50m·min^{-1} 84

Figure 4.45 Filament diameter vs. distance for the spinning conditions: v= 50m·min^{-1}; Q= 0.5, 1.0, 1.5, and 2.0 g·min^{-1} 84

Figure 4.46 Filament diameter vs. distance for the spinning conditions: Q= 1.0 g·min^{-1}; v=10, 30, 50, and 70 m·min^{-1} 84

Figure 4.47 Air drag and inertial force at a distance 180 cm from the spinneret for different spinning conditions: (a) different mass flow rates and constant take-up velocity, (b) different take-up velocities and constant mass flow rate.. 86

Figure 4.48 Tensile force vs. distance for the spinning condition A: Vertical y-axis from 0.0 to 2.5 (a), and 1.90 to 2.04 (b) 87

Figure 4.49 Tensile force vs. distance for the spinning condition B: Vertical y-axis from 0.0 to 2.5 (a), and 1.90 to 2.10 (b) 87

Figure 4.50 Tensile stress vs. distance for the spinning condition A (a) and B (b)............................. 88

Figure 4.51 Tensile stress at maximum ASR for different spinning conditions A (a) and B (b) 88

List of figures

Figure 4.52 Apparent elongational viscosity vs. distance for the spinning condition A (a and c) and spinning condition B (b and d) 90

Figure 4.53 Solidification distance from fitted temperature profiles LST (the red-coloured filled circle (●)) and fitted velocity profiles LSv (the black-coloured filled square (■)) vs. mass flow rate (a) and vs. take-up velocity (b) 91

Figure 4.54 Solidification time vs. mass flow rate (a) and take-up velocity (b) 92

Figure 4.55 Schematic view of a monofilament and locations of the captured samples 93

Figure 4.56 SEM images of the dispersed PLA phase from PLA/PVAL blend granules (P0), prepared using internal mixer after removing the PVAL matrix: (a) scale bar 10 µm; (b) scale bar: 1 µm 94

Figure 4.57 SEM images of the dispersed PLA phase from PLA/PVAL blend granules (P0), prepared using twin-screw extruder after removing the PVAL matrix: (a) scale bar 10 µm; (b) scale bar: 1 µm 94

Figure 4.58 SEM images of the dispersed PLA phase from PLA/PVAL extrudates at the location near the die exit (P1) after removing the PVAL matrix: (a) scale bar: 10µm; (b) scale bar: 1µm 95

Figure 4.59 SEM images of PLA nanofibers from PLA/PVAL blend filaments (P8) after removing the PVAL matrix, spinning speed: 50 m·min^{-1}: (a) scale bar: 10µm; (b) scale bar: 1µm; (c) scale bar: 200 nm 95

Figure 4.60 SEM images of PLA structures (left and middle columns) after removing PVAL matrix and PLA/PVAL cross-section after etching dispersed PLA phases (right column) at different locations along the spinline: P2 (a — c), P3 (d — f), P4 (g — i), P5 (J—L), P6 (m — o), and P7 (p — r). 100

Figure 4.61 A conceptual model of the mechanisms of fibrillation process of PLA/PVAL blends in elongational flow along the spinnline under special spinning conditions $D0$= 0.6 mm with $L/D0$=2, Q=1.0 g·min^{-1} (V=0.78 cm^3·min^{-1}), v=50 m· min^{-1}, T=195 °C 101

Figure 4.62 Schematic drawing three representative possible models of molten fiber deformation on the very first distances just below the die exit (x ≈ 0 — 3 cm) at symmetrical radial viscosity distribution: (a) symmetrical radial velocity distribution and uniform radial stress distribution, (b) symmetrical radial velocity and stress distribution, (c) uniform radial velocity distribution and symmetrical stress distribution, modified after [146, 175] 102

Figure 4.63 SEM images of the PLA/PVAL 30/70 blend pellets after removing PVAL matrix (a–c) and etching PLA dispersed phase (d–f) with scale bars for the left, middle, and right column are 10, 1, and 1 µm, respectively. 105

Figure 4.64 Frequency-distribution histograms vs. circular equivalent diameter (CED) over the range of CED up to 0.5 µm (a) and CED from 0.5 to 13 µm (b) 105

Figure 4.65 Cumulative number percentage vs. CED (a and b) and log-normal distributions (c) 106

List of figures

Figure 4.66 Comparison of the maximum diameter of PLA droplet vs. shear rate range: (a) linear scale and (b) logarithmic scale for the y-axis. 106

Figure 4.67 Morphology of the cross-sectional surfaces of PLA/PVAL blend extrudates after etching dispersed PLA phase for the various mass flow rates (extrusion rate) of 0.5 (a–c), 1.0 (d–f), 1.5 (g–i), and 2.0 g·min^{-1} (j–l): The scale bars from the left to right column are 10, 1, and 1 μm 107

Figure 4.68 Morphology of dispersed PLA phase after removing PVAL matrix for the various mass flow rates (extrusion rate) of 0.5 (a–c), 1.0 (d–f), 1.5 (g–i), and 2.0 g·min^{-1} (j–l) with scale bars from the left to right column are 10 μm, 1 μm, and 1 μm, respectively. 108

Figure 4.69 Maximum and number average CED $dCED$ vs. mass flow rate for the cross-section of PLA/PVAL blend extrudates after etching dispersed PLA phase (measured from Figure 4.67) 108

Figure 4.70 Cumulative number vs. CED (a), Cumulative number percentage vs. CED (b) 109

Figure 4.71 Measured pressure for various mass flow rates (a) and total pressure acting on the polymer blends in a convergent capillary die (b): the red colour droplets represent the very small droplets. There has been almost no coalescence among these small droplets when they passing though convergent capillary die. 111

Figure 4.72 Mean diameters d with deviations of the dispersed PLA phase from PLA/PVAL blend extrudates after removing the PVAL matrix for various mass flow rates (measured from Figure 4.68). 112

Figure 4.73 Schematic presentation of coalescence and deformation of droplets and different possible diameters of an ellipsoidal PLA droplet in PLA/PVAL blend extrudates. 112

Figure 4.74 SEM images of dispersed PLA phase after removing PVAL matrix for the various mass flow rates (0.5 – 2.0 g·min^{-1}) (Q05 – Q20) and the constant take-up velocity of 50 m·min^{-1} (V50) (Q05V50, Q10V50, Q15V50, Q20V50) at different locations (P1 – P7) along the spinline: scale bar is 1 μm, (*) Experiments were not done at this location because it was supposed that there is no difference in PLA morphology at this location with that at x=30 cm (P6). 114

Figure 4.75 SEM images of dispersed PLA phase after removing PVAL matrix for various take-up velocities (10 – 70 m·min^{-1}) (V10 – 70) and the constant mass flow rate of 1.0 g·min^{-1} (Q10) (V10Q10, V30Q10, V50Q10, V70Q10) at different locations (P1 – P7) along the spinline: scale bar is 1 μm, (*) Experiments were not done at this location because it was supposed that there is no difference in PLA morphology at this location with that at x=30 cm (P6) 115

Figure 4.76 Mean diameter d of dispersed PLA phase after removing PVAL matrix vs. distance ... 116

Figure 4.77 Maximum ASR and the position of maximum ASR vs. mass flow rate (a) and take-up velocity (b) 117

List of figures

Figure 4.78 Temperature and apparent elongational viscosity of filament at maximum ASR vs. mass flow rate (a) and take-up velocity (b) 117

Figure 4.79 Average diameter of PLA fibrils at P6 (x=30 cm) $dx30$, P7 (x=50 cm) $dx50$, and P8 $dL(xL = 200$ cm) vs. mass flow rate (a) and take-up velocity (b) 119

Figure 4.80 SEM images of dispersed PLA phase from PLA/PVAL blend extrudates at P7 (left) and PLA/PVAL blend filaments at P8 (right) for the two special spinning conditions: Q=2.0 g·min^{-1} and v=50 m·min^{-1} (a and b), v=10 m·min^{-1} and Q=1.0 g·min^{-1} (c and d). Scale bar: 1 μm 120

Figure 4.81 SEM images of dispersed PLA phase from PLA/PVAL blend extrudates at P6 (left) and PLA/PVAL blend filaments at P8 (right) for the two limiting spinning conditions: Q=0.5 g·min^{-1} and v=50 m·min^{-1} (a and b), v=70 m·min^{-1} and Q=1.0 g·min^{-1} (c and d). Scale bar: 1 μm 120

Figure 4.82 SEM images of dispersed PLA phase from PLA/PVAL blend extrudates at P7 (left) and from PLA/PVAL blend filaments at P8 (right) for the three last spinning conditions: Q=1.0 g·min^{-1} and v=30 m·min^{-1} (a and b); Q=1.0 g·min^{-1} and v=50 m·min^{-1} (c and d); Q=1.5 g·min^{-1} and v=50 m·min^{-1} (e and f). Scale bar: 1 μm 121

Figure 4.83 SEM images of cross-sectional PLA/PVAL blend extrudates after etching the dispersed PLA phase for the various mass flow rates (0.5 – 2.0 g·min^{-1}) (Q05 – Q20) and the constant take-up velocity of 50 m·min^{-1} (V50) (Q05V50, Q10V50, Q15V50, Q20V50) at different locations (P1 – P7) along the spinline: scale bar 1 μm, (*) Experiments were not done at this location because it was supposed that there is no difference in PLA morphology at this location with that at x=20 cm (P5). 124

Figure 4.84 SEM images of cross-sectional PLA/PVAL blend extrudates after etching the dispersed PLA phase for various take-up velocities (10 – 70 m·min^{-1}) (V10 – V70) and the constant mass flow rate of 1.0 g·min^{-1} (Q10) (V10Q10, V30Q10, V50Q10, V70Q10) at different locations (P1 – P7) along the spinline: scale bar 1 μm, (*) Experiments were not done at this location because it was supposed that there is no difference in PLA morphology at this location with that at x=30 cm (P6) 125

Figure 4.85 Mean CED of the dispersed PLA phase in cross-sectional PLA/PVAL blend extrudates after etching the PLA phase vs. distance to spinneret 126

Figure 4.86 A schematic draw of a possible radial coalescence of neighbour PLA droplets in a PLA/PVAL blend extrudate during stretching process under effect of elongational and compressional stresses 126

Figure 4.87 SEM images of PLA/PVAL blend extrudates at position P6 (x=30 cm) after removing PVAL matrix (right column) and etching dispersed PLA phase (left column) for the three different spinning conditions: Q=1.0 g·min^{-1} and v=50 m·min^{-1} (Q10V50); Q=1.5 g·min^{-1} and v=50 m·min^{-1} (Q15V50); Q=1.0 g·min^{-1} and v=30 m·min^{-1} (Q10V30), scale bar: 1 μm 127

List of figures

Figure 4.88 Mean diameter d and mean CED $dCED$ of the dispersed PLA phase from PLA/PVAL blend extrudates at position P6 (x=30 cm) for the three different spinning conditions: Q=1.0 g·min^{-1} and v=50 m·min^{-1} (Q10V50); Q=1.5 g·min^{-1} and v=50 m·min^{-1} (Q15V50); Q=1.0 g·min^{-1} and v=30 m·min^{-1} (Q10V30) 127

Figure 4.89 Cumulative number percentage vs. diameter d/CED $dCED$ of dispersed PLA phase from PLA/PVAL blend extrudates at position P6 (x=30 cm) for the last three different spinning conditions: (a) Q=1.0 g·min^{-1} and v=50 m·min^{-1} (Q10V50); (b) Q=1.5 g·min^{-1} and v=50 m·min^{-1} (Q15V50); (c) Q=1.0 g·min^{-1} and v=30 m·min^{-1} (Q10V30). (d) Cumulative number percentage of diameter d/CED $dCED$ having diameter up to 0.1 µm 128

Figure 4.90 An overview of possible conceptual models of the deformation, coalescence, and break-up processes for different spinning conditions 130

Figure 4.91 A relative scale of the deformation levels. (*) The term "deformation (De)" does not mean that its deformation level in Group I is the same value as its deformation level in Group II or III. All the terms "Almost no De", "Small De", "De", "Large De", and "Largest De" are used to compare the deformation level within each group. 131

List of tables

Table 2.1 Effect of hydrolysis degrees and molecular weight on the physical properties of PVAL, modified after [83]..................9

Table 2.2 Nusselt vs. Reynolds number relations, modified after [175] 25

Table 3.1 PVAL Specialties 27

Table 3.2 PLA Specialties 28

Table 4.1 The glass transition temperature of PLA, PVAL, and their blends 49

Table 4.2 Main ATR-FTIR vibrational bands in PVAL (88 % hydrolyzed) and in PLA 52

Table 4.3 Textile-physical properties of melt spun/offline-drawn PLA/PVAL filaments 66

Table 4.4 Spinning conditions with an extrusion temperature of 195 °C, D_0=0.6 mm, L/D=2..................77

Table 4.5 Tensile stress at maximum axial strain rate (ASR) for different spinning conditions..................88

Table 4.6 Maximum axial strain rate (ASR) and its locations for different spinning conditions..................116

Table 4.7 Average diameters d of PLA fibrils after removing PVAL matrix at locations P6 dx30 (x=30 cm), P7 dx50 (x=50 cm), and P8 dL (x=200 cm) 119

Table 4.8 An overview of the possibility of the deformation (De), coalescence (Co), and break-up (Bu) processes for different spinning conditions..................129

Table 4.9 Different possible sequences of droplet deformation, coalescence, and break-up 131

Appendix A
Experimental and simulated temperature profiles of PLA/PVAL monofilament

The Appendix A is divided into two main sections starting with experimental results of the temperature variations of PLA/PVAL monofilament along the spinline under different spinning conditions as shown in Table 4.4 (chapter 4, section 4.3). The exponential fitting curves obtained from the experimental results were used to determine the Nusselt number. In the next section, the above obtained Nusselt numbers were used to simulate the filament temperature profiles. The simulated results were then compared to common model of melt spinning process developed by Kase und Matsuo.

A.1 Experimental filament temperature profiles and Nusselt number

A.1.1 Experimental results of filament temperature profiles

Figure A.1 and A.2 show the experimental results of filament temperature under the different spinning conditions. The grey cross symbols (×) present for the filament temperature values obtained using an Infrared camera as mentioned in chapter 3 (section 3.3.6). Points (including deviation) are the corrected temperatures by emissivity and the dash curves (— — —) are the temperature curves fitted by using different exponential decay functions as follow:

$$T(x) = T_{air} + (T_0 - T_{air}) \times \exp\left(\frac{-x}{u}\right) \qquad (A.1)$$

where T_{air}=25 °C and T_0=195 °C are the temperature of the ambient air and the temperature of molten filament at die exit (x=0), respectively, x is the distance from the spinneret, and different constant u values are directly presented in the Figure A.1 and A.2.

Appendix A

A.1.2 Nusselt numbers Nu

The Nusselt number Nu was determined using the following Equation A.2, is reformulated from the Equation 2.29 (section 2.3.3) with the assumption that the PLA/PVAL filament may not crystallize during melt spinning of the low spinning speed:

$$N_u = -\frac{dT}{dx} \cdot \frac{1}{T(x) - T_{air}} \cdot \frac{Q \cdot c_p}{\pi \cdot \lambda_{air}} \qquad (A.2)$$

where, $T(x)$ are the fitted temperature profiles obtained using the above Equation A.1.

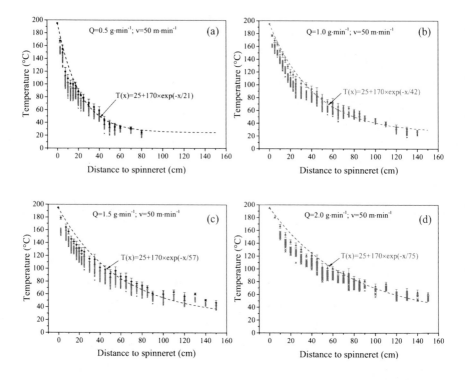

Figure A.1 Filament temperature vs. distance for the different mass flow rate of 0.5 g·min^{-1} (a), 1.0 g·min^{-1} (b), 1.5 g·min^{-1} (c), and 2.0 g·min^{-1} (d) at the take-up velocity of 50 m·min^{-1}

Appendix A

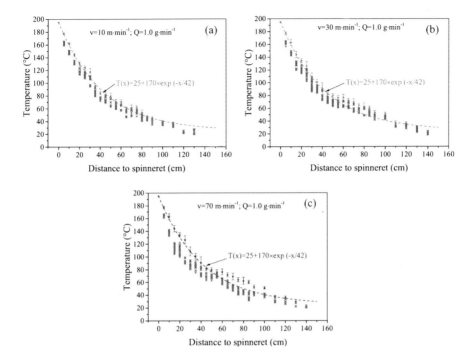

Figure A.2 Filament temperature vs. distance for the mass flow rate of 1.0 g·min^{-1} at the different take-up velocity of 10 m·min^{-1}(a), 30 m·min^{-1}(b), and 70 m·min^{-1}(c)

Figure A.3 shows the relationship between the Nusselt number and temperature as well as the Nusselt number and Reynolds number. It can be seen that the Nusselt number is nearly constant along spinline. The Nusselt number has higher values at the higher amount of mass flow rates. However, there is little difference between the Nusselt number of the mass flow rates of 0.5, 1.0 g·min^{-1} and that of the mass flow rates of 1.5, 2.0 g·min^{-1} at the constant take-up velocity of 50 m·min^{-1}. For the constant mass flow rate of 1.0 g·min^{-1}, the Nusselt number is also nearly constant and has the same value of 0.85 at the different velocities up to 70 m·min^{-1}. In this study, the Nusselt number is approximately considered as a constant value of 0.85 if $Q \leq 1.0$ g·min^{-1} and 0.95 if $1.0 < Q \leq 2.0$ g·min^{-1} independent of take-up velocities within the above range of Reynolds number.

Appendix A

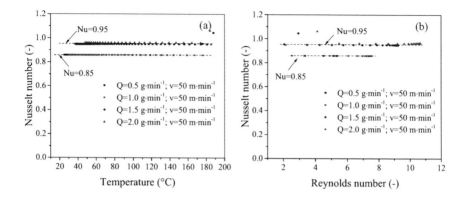

Figure A.3 Nusselt number vs. temperature (a) and Nusselt number vs. Reynolds number (b) for the different mass flow rates at the take-up velocity of 50 m·min⁻¹

A.2 Simulated temperature profiles of PLA/PVAL monofilament

A.2.1 Spinning parameters

The temperature $T(x)$ of a running PLA/PVAL filament along the spinline was simulated using the following Equation A.3

$$\frac{dT(x)}{dx} = -(T(x) - T_{air}(x)) \cdot N_u \frac{\pi \cdot \lambda_{air}}{Q \cdot c_p} \tag{A.3}$$

where all spinning parameters are summarized in Table A.1 below.

Table A.1 Spinning and material parameters for the simulation of filament temperature along the spinline

Spinning parameters (unit)	Value
- The temperature of surrounding air T_{air} (°C)	25
- The heat conductivity of air at 25 °C λ_{air} ($W \cdot m^{-1} \cdot K^{-1}$)	0.026
- The mass density of the polymer blend ϱ_P ($g \cdot cm^{-3}$)	1.25
- The specific heat capacity of the PLA/PVAL 30/70 blend c_p ($J \cdot g^{-1} \cdot K^{-1}$)	1.80 [*]
- Nusselt number N_u (-)	0.85/0.95 [**]

[*] determined using the differential scanning calorimetry (DSC) measurement

[**] For different mass flow rates (Nu=0.85 if Q ≤ 1.0 g·min⁻¹; Nu=0.95 if 1.0 < Q ≤ 2.0 g·min⁻¹)

Appendix A

A.2.2 Comparison the measured filament temperature profiles with simulated results as well as with Kase and Matsuo model

The filament temperature profiles of both measured and simulated results were compared with the Kase and Matsuo model that was calculated using the Equation A.3 with the Nusselt number equation $N_u = 0.42 \cdot Re_\parallel^{0.334}$ for air flowing parallel to a cylinder.

Figure A.4 and A.5 show again the experimental results and their comparison with by our simulated results as well as with Kase and Matsuo model. The dash curves (— — —) are the fitted temperature curves, the continuous curves (———) are the simulated curves, and the dash dot curves (— · · — · · —) present the filament temperature profiles obtained using the Kase and Matsuo model. It can be seen that the simulated results agreed well with measured values of filament temperatures. The simulated PLA/PVAL filament temperature cooled faster than the PLA/PVAL filament temperature obtained using the Kase and Matsuo model.

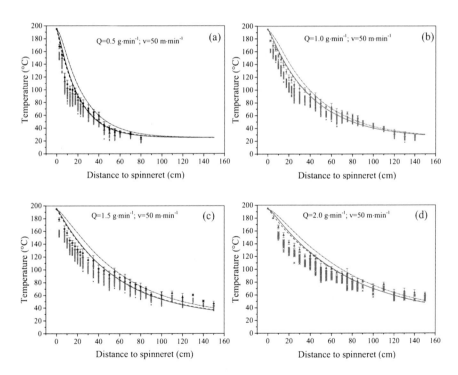

Figure A.4 Filament temperature vs. distance for the different mass flow rate of 0.5 g·min⁻¹ (a), 1.0 g·min⁻¹ (b), 1.5 g·min⁻¹ (c), and 2.0 g·min⁻¹ (d) at the take-up velocity of 50 m·min⁻¹

Appendix A

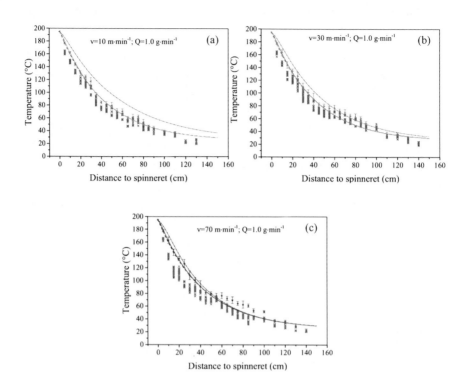

Figure A.5 Filament temperature vs. distance for the mass flow rate of 1.0 g·min^{-1} at the different take-up velocity of 10 m·min^{-1} (a), 30 m·min^{-1} (b), and 70 m·min^{-1} (c)

Appendix B
Experimental results of the filament velocity profiles of PLA/PVAL monofilament

The Appendix B provides experimental results of filament velocity profiles along the spinline under different spinning conditions as shown in Table 4.4 (chapter 4, section 4.3). The grey cross symbols (×) present for the velocity values obtained using a laser Doppler anemometry device and the spline-connected curves are the fitted curves by manual adjustment.

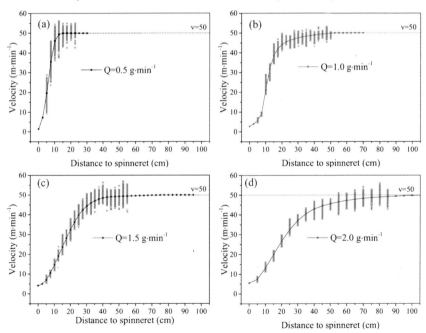

Figure B.1 Filament velocity vs. distance for the different mass flow rates of 0.5 g·min^{-1}(a), 1.0 g·min^{-1}(b), 1.5 g·min^{-1}(c), and 2.0 g·min^{-1}(d) at the take-up velocity of 50 m·min^{-1}

Appendix B

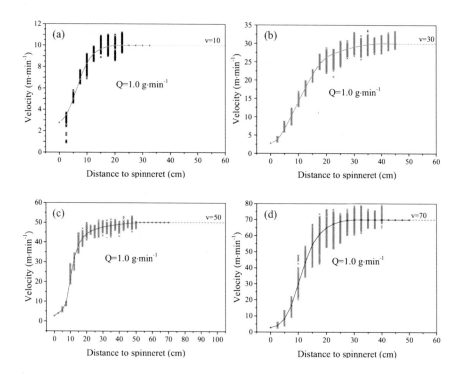

Figure B.2 Filament velocity vs. distance for the mass flow rate of 1.0 g·min^{-1} at the different take-up velocities of 10 m·min^{-1} (a), 30 m·min^{-1} (b), 50 m·min^{-1} (c), and 70 m·min^{-1} (d)

Appendix C
Original SEM images

The Appendix C provides selected original SEM images of PLA/PVAL 30/70 blends after removing either PLA or PVAL component. The prepared PLA/PVAL blend samples for SEM investigation were presented in Figure 4.55. They include blend granules before extrusion and melt spinning, blend extrudates without stretching, and captured blend extrudates at different locations along the spinline.

The abbreviation of blend samples is described as follows:

$$XijQklVmn$$

It includes 3 parts. First one **Xij** describes positions along the spinline, where the filament taken for investigation of morphology. For examples, X00, X01, X05, and X10 are presented for the positions x=0 cm, x=1 cm, x=5, and x=10 cm from the die exit, respectively. Middle part **Qkl** describes mass flow rates. They are Q05, Q10, Q15, and Q20, which are presented for various mass flow rates, Q=0.5, 1.0, 1.5, and 2.0 g·min^{-1}, respectively. Last one **Vmn** represents take-up velocity. It includes V10, V30, V50, and V70 presenting for different take-up velocities, v=10, 30, 50, and 70 m·min^{-1}, respectively.

If one reads **X01Q10V50** that means the samples are taken at position x=1 cm below the die exit for the mass flow rate of 1.0 g·min^{-1} and the take-up velocity of 50 m·min^{-1}.

C.1 Morphology of PLA/PVAL blend granules

Figure C.1 presents the original SEM images of the PLA/PVAL 30/70 blend pellets obtained from the twin-screw extruder. Scale bars are 10, 1, and 1 μm from top to bottom, respectively.

Appendix C

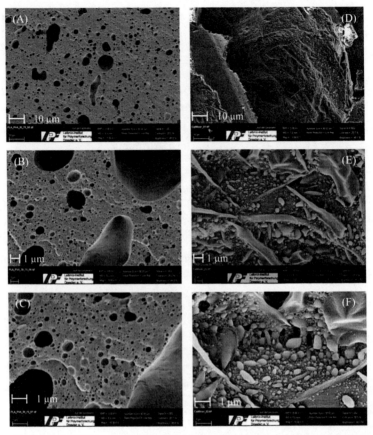

Figure C.1 Original SEM images of PLA/PVAL blend pellets after etching PLA dispersed phase (A – C, left column) and removing PVAL matrix (D – F, right column) and with scale bars 10, 1, and 1 μm from top to bottom, respectively

C.2 Morphology of PLA/PVAL blend samples at the initial position (x=0)

Figure C.2 represents the selected original SEM images of the PLA/PVAL blend extrudates without stretching (as known initial position x=0) of various mass flow rates 0.5 – 2.0 g·min^{-1}. The scale bar is 1 μm, corresponds to 10.000× magnification.

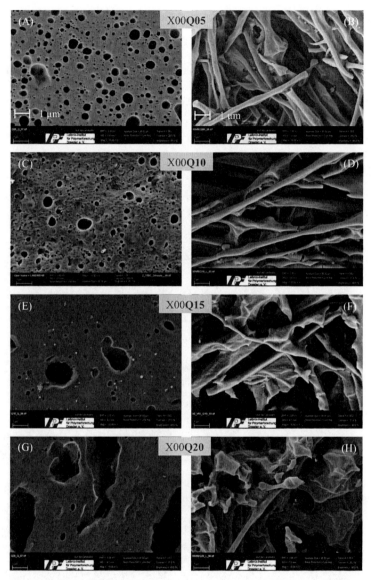

Figure C.2 Original SEM images of the PLA/PVAL blend extrdates for the various flow rates of 0.5 g·min^{-1} (A and B), 1.0 g·min^{-1} (C and D), 1.5 g·min^{-1} (E and F), and 2.0 g·min^{-1} (G and H). The scale bar is 1 μm.

Appendix C

C.3 Morphology of blend filament at different locations along the spinline

It is worth noting that the scale bar is 1 μm for all the SEM images presented below.

C.3.1 For the spinning condition A: constant take-up velocity and various flow rates

Figure C.3 — C.6 represent for the original SEM images of the PLA/PVAL blend extrudates at different locations along the spinline for the various mass flow rates of Q=0.5, 1.0, 1.5, and 2.0 g·min^{-1} with the constant take-up velocity of v=50 m·min^{-1}.

Mass flow rate Q=0.5 g·min^{-1} and take-up velocity v=50 m·min^{-1}

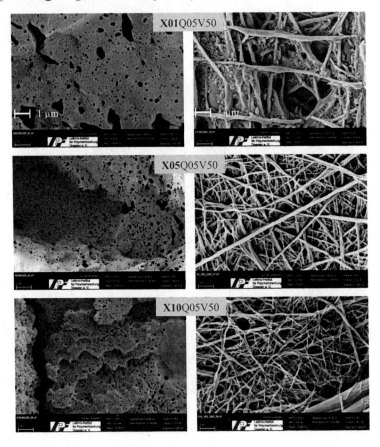

Appendix C

Figure C.3 Original SEM images of the PLA/PVAL blend extrudates at different location along the spinline for the mass flow rate of 0.5 g·min^{-1} and the take-up velocity of 50 m·min^{-1}

Mass flow rate Q=1.0 g·min-1 and take-up velocity v=50 m·min^{-1}

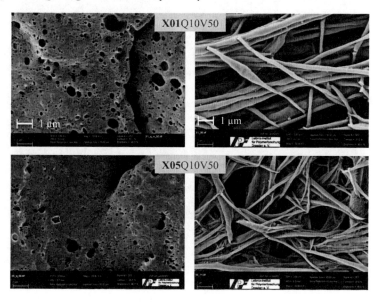

Appendix C

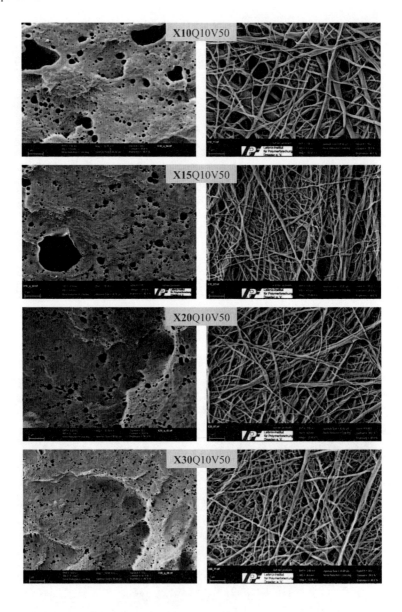

C—6

Appendix C

Figure C.4 Original SEM images of the PLA/PVAL blend extrudates at different location along the spinline for the mass flow rate of 1.0 g·min^{-1} and the take-up velocity of 50 m·min^{-1}

Mass flow rate Q=1.5 g·min^{-1} and take-up velocity v=50 m·min^{-1}

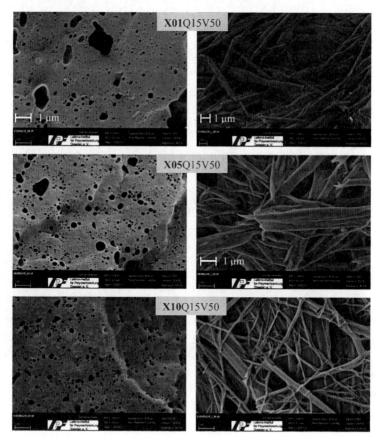

Appendix C

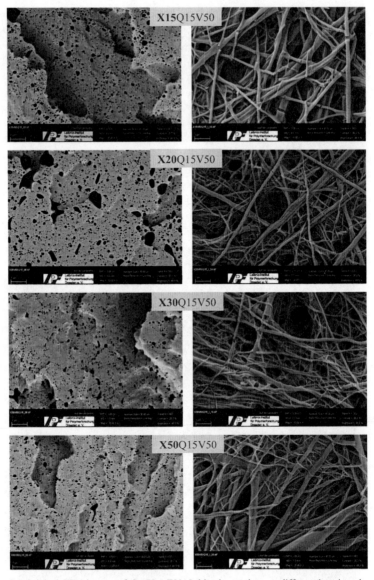

Figure C.5 Original SEM images of the PLA/PVAL blend extrudates at different location along the spinline for the mass flow rate of 1.5 g·min^{-1} and the take-up velocity of 50 m·min^{-1}

Appendix C

Mass flow rate Q=2.0 g·min^{-1} and take-up velocity v=50 m·min^{-1}

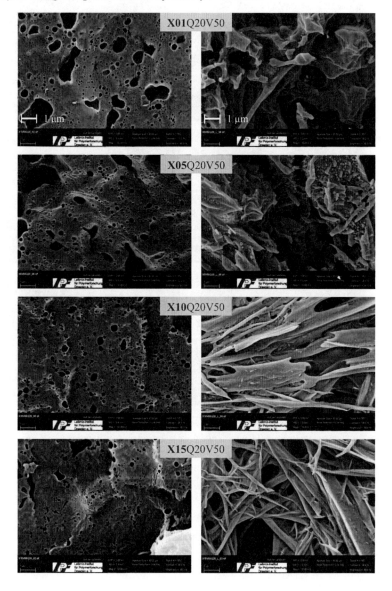

C—9

Appendix C

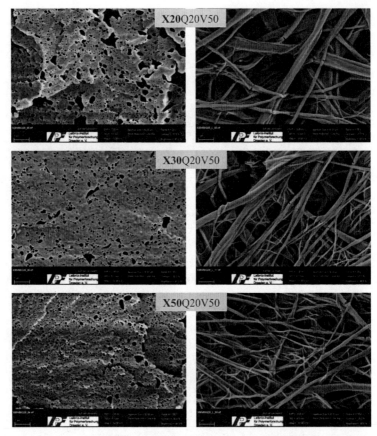

Figure C.6 Original SEM images of the PLA/PVAL blend extrudates at different location along the spinline for the mass flow rate of 2.0 g·min^{-1} and the take-up velocity of 50 m·min^{-1}

C.3.2 For the spinning condition B: constant flow rate and various take-up velocities

Figure C.7 – C.9 represent for the original SEM images the PLA/PVAL blend extrudates at different locations along the spinline for the constant flow rate of Q=1.0 g·min^{-1} with the various take-up velocities of 10, 30, and 70 m·min^{-1}. As mentioned above, the SEM images on the left column are the cross sections of PLA/PVAL blend samples after etching the dispersed PLA phase. The PLA dispersed domains are observed as black holes in the PVAL matrix. The SEM images on the right column show the remaining dispersed PLA phase after removing the PVAL matrix.

Appendix C

Take-up velocity v=10 m·min^{-1} and mass flow rate Q=1.0 g·min^{-1}

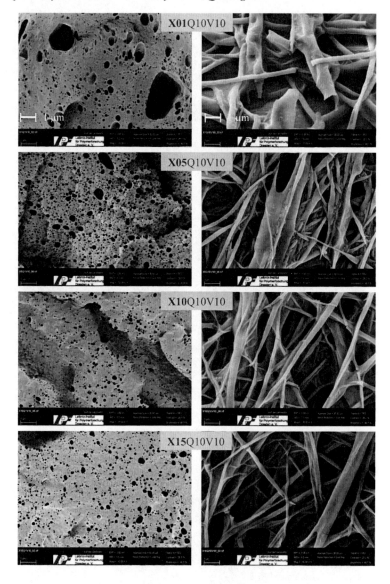

C−11

Appendix C

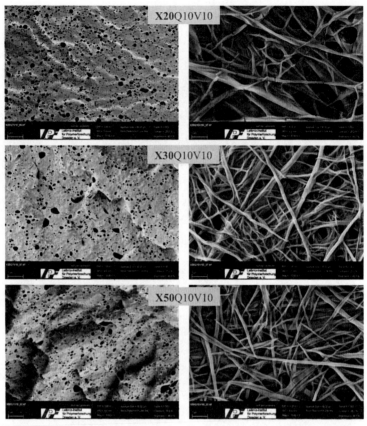

Figure C.7 Original SEM images of the PLA/PVAL blend extrudates at different location along the spinline for the mass flow rate of Q=1.0 g·min^{-1} and the take-up velocity of 10 m·min^{-1}

Take-up velocity v=30 m·min^{-1} and mass flow rate Q=1.0 g·min^{-1}

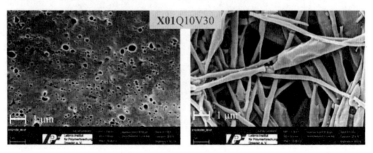

Appendix C

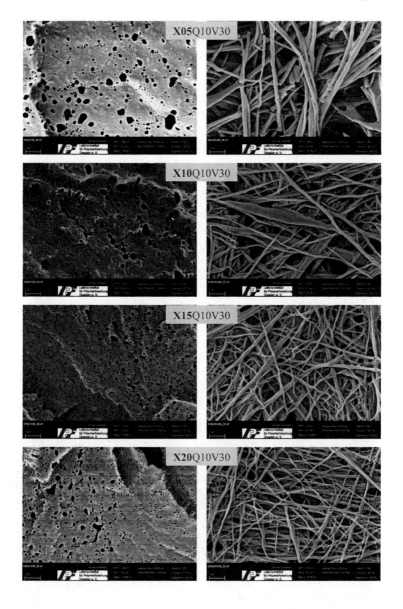

C—13

Appendix C

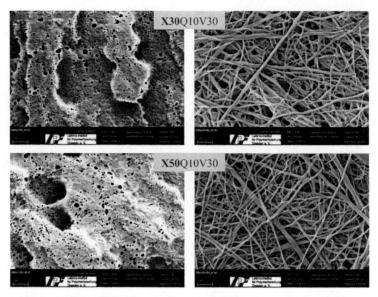

Figure C.8 Original SEM images of the PLA/PVAL blend extrudates at different location along the spinline for the mass flow rate of Q=1.0 g·min^{-1} and the take-up velocity of 30 m·min^{-1}

Take-up velocity v=70 m·min^{-1} and mass flow rate Q=1.0 g·min^{-1}

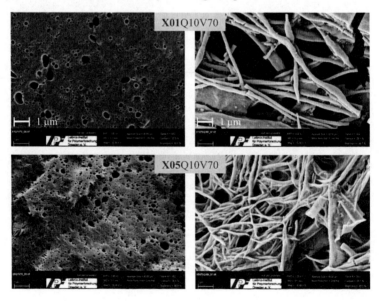

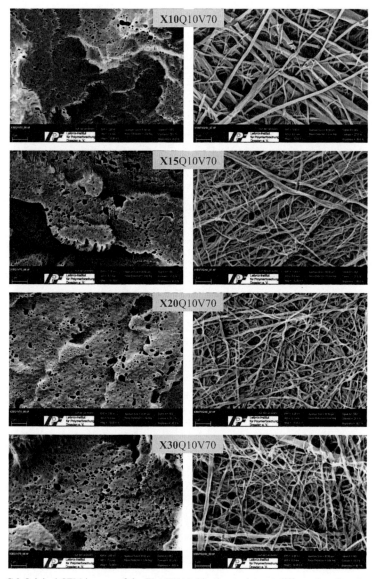

Figure C.9 Original SEM images of the PLA/PVAL blend extrudates at different location along the spinline for the mass flow rate of Q=1.0 g·min^{-1} and the take-up velocity of 70 m·min^{-1}

Appendix D
Others

The Appendix D provides theoretical calculation of the interfacial tension between PLA and PVAL (Section D.1). It presents also calculation of the deformation of PLA droplets in PLA/PVAL blends according to affine theory (Section D.2).

D.1 Interfacial tension between PLA and PVAL

Because the difficulties in determining the interfacial tension by experiment, the interfacial tension $\Gamma_{PLA,PVAL}$ between poly(lactic acid) (PLA) and poly(vinyl alcohol) (PVAL) in this study was estimated by theoretical calculation using the following equation [1][1]

$$\Gamma_{PLA,PVAL} = \Gamma_{PLA} + \Gamma_{PVAL} - 2\phi\sqrt{\Gamma_{PLA}\Gamma_{PVAL}} \qquad (D.1)$$

where Γ_{PLA} and Γ_{PVAL} are the surface tension of PLA and PVAL, respectively. The interaction parameters ϕ is simply assumed equals 1 ($\phi = 1$).

D.1.1 Surface tension of PLA Γ_{PLA}

The surface tension of PLA Γ_{PLA}=37.5 mN·m^{-1} was calculated from the Parachor P_s was developed by Quayle [2]

$$\Gamma_{PLA} = \left(\frac{P_{s,PLA}}{V}\right)^4 \qquad (D.2)$$

where $V = M/\rho$ is the molar volume of PLA, in which ρ_{PLA}=1.24 g·cm^{-3} and M=72.06 are the mass density and the molar mass of PLA, respectively. The molar Parachor of PLA

---
[1] References can be found at the end of the Appendix D

Appendix D

$P_{s,PLA}$=143.8 was determined from the group contributions as presented by Quayle are given in table D.1

Table D.1 Atomic and structural contributions to the parachor P_s assigned by Quayle [2]

Unit	C	H	CH_2	O	O_2 (in esters)
Values assigned by Quayle	9.0	15.5	40.0	19.8	54.8

D.1.2 Surface tension of PVAL Γ_{PVAL}

It should be remembered from the section 4.1 that the used PVAL in this study is the partially PVAL hydrolyzed 88 % was prepared by the saponification of poly (vinyl acetate) (PVAc) having vinyl alcohol and vinyl acetate groups. This means that the chemical structure of PVAL includes both the hydroxyl groups (-OH) and carbonyl (-OCOCH$_3$), as shown in Figure D.1, called poly (vinyl alcohol-co-vinyl acetate). The PVAL hydrolyzed 88 % means that poly (vinyl alcohol-co-vinyl acetate) has 88 % mole percent of pure poly (vinyl alcohol) [-CH(OH)-CH$_2$)-]m (PVA) and 12 % mole percent of pure poly (vinyl acetate) [-CH$_2$-CH(OCOCH$_3$)-]n (PVAc).

Figure D.1 Chemical structures of poly (vinyl alcohol-co-vinyl acetate)

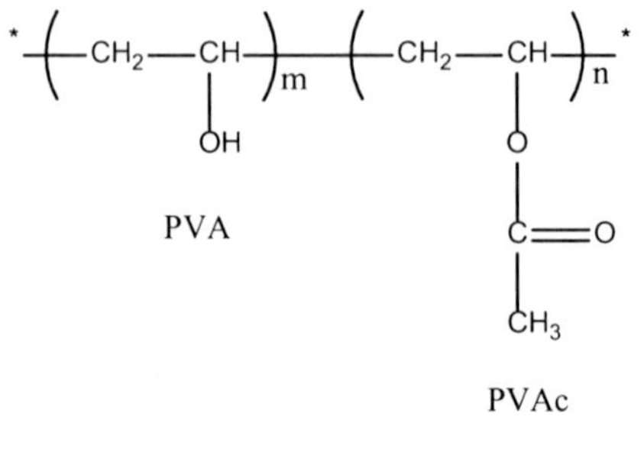

Like the calculation of the PLA surface tension Γ_{PLA}, the surface tension of PVA Γ_{PVA} and PVAc Γ_{PVAc} have values of 59 and 40 mN·m^{-1}, respectively [3].

Table D.2 summarizes the calculation of volume percent of PVA and PVAc for 1 mole PVAL. In 1 mole PVAL has 32.7 % volume PVA and 67.3 % mole PVAc. If 1 mole PVAL includes 88 % mole PVA and 12 % PVAc, it has 78 % volume PVA and 22 % volume PVAc.

Then, the surface tension of PVAL Γ_{PVAL} can be calculated as follows:

Appendix D

$$\Gamma_{PVAL} = \Gamma_{PVA} \times vol\ \%\ PVA + \Gamma_{PVAc} \times vol\ \%\ PVAc \quad (D.3)$$

$$\Gamma_{PVAL} = \Gamma_{PVA} \times 78\ \% + \Gamma_{PVAc} \times 22\ \% = 54.8 \quad (D.4)$$

Using the Equation (D.1) with Γ_{PLA}=37.5 mN·m^{-1}, Γ_{PVAL}=54.8 mN·m^{-1}, and $\phi = 1$, the interfacial tension between PLA and PVAL is now 1.64 mN·m^{-1}.

Table D.2 Volume percent of PVA and PVAc in 1 mole PVAL

	Mass density (g·cm^{-3})	Molar mass (g·mol^{-1})	Molar volume (cm^3·mol^{-1})	% volume
PVA	1.26	44.2	35.08	32.66
PVAc	1.19	88.06	72.34	67.34

D.2 Deformation of PLA droplet

Figure D.2 present comparison between the measured diameter of PLA phase after removing PVAL matrix d and the calculated diameter of PLA phase using the affine deformation theory B after Equation 2.12, in section 2.2.4, as rewritten as follows:

$$B_{x+1} = B_x exp\left(-\frac{\dot{\varepsilon}}{2}\delta t\right) \quad (D.5)$$

where, $\dot{\varepsilon}$ is the local average elongation rate $\dot{\varepsilon} = \delta v(x)/\delta x$ and δt is the period of time within δx= 2.5 cm along the spinline.

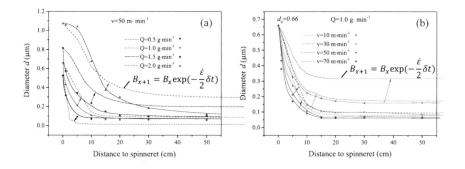

Figure D.2 Comparison between calculated diameter of PLA droplets/fibrils after affine deformation theory B and measured diameters of PLA phase after removing PVAL matrix d: (a) v=50 m·min^{-1}; Q= 0.5, 1.0, 1.5, and 2.0 g·min^{-1}; (b) Q=1.0 g·min^{-1}; v= 10, 30, 50, and 70 m·min^{-1}

D–3

Appendix D

It can be clearly seen that the calculated diameter of PLA phase using the affine deformation theory B and the measured diameter of PLA phase after removing PVAL matrix d could not be well fitted together.

In principle, the calculation of PLA droplet deformation was calculated using the affine deformation theory with an assumption that PLA droplets/fibrils have cylindrical shapes with their volume being conserved in non-isothermal melt spinning process. Furthermore, the affine deformation theory is only appropriate for small deformations and for a few droplets, in which all droplets have the same droplet size.

Therefore, it can be concluded that the affine theory are not applicable for the PLA droplet deformation in PLA/PVAL blend extrudates along the spinline, because the distribution of PLA droplets is very broad. All the droplets could not be counted during calculating using the affine deformation theory.

References for Appendix D

[1] Girifalco, L.A. and Good R.J., A Theory for the Estimation of Surface and Interfacial Energies. I. Derivation and Application to Interfacial Tension. The Journal of Physical Chemistry **61**(7), 904-909 **(1957)**

[2] Quayle, O.R., The Parachors of Organic Compounds. An Interpretation and Catalogue. Chemical Reviews **53**(3), 439-589 **(1953)**

[3] Van Krevelen D.W. and Nijenhuis K.T., in Properties of Polymers (Fourth Edition), Van Krevelen D.W. and Nijenhuis K.T., Eds, Elsevier: Amsterdam **2009**.

List of publications

Journal Articles

1. N.H.A. Tran, H. Brünig, C. Hinüber, G. Heinrich. Melt spinning of nano fibrillary structures from poly (lactic acid) and poly (vinyl alcohol) blends. Macromol. Mater. Eng. **2014**. 299(2): p. 219-227
2. N.H.A. Tran, H. Brünig, R. Boldt, G. Heinrich. Morphology development from rod-like to nanofibrillar structures of dispersed poly (lactic acid) phase in a binary blend with poly (vinyl alcohol) matrix along the spinline. Polymer **2014**. 55(24): p. 6354-6363
3. N.H.A. Tran, H. Brünig, G. Heinrich. Characterization of filament profiles in the low-speed melt spinning of poly(lactic acid) and poly(vinyl alcohol) blend (*in preparation*)
4. N.H.A. Tran, H. Brünig, M. Auf der Landwehr, G. Heinrich. Controlling the micro-and nanofibrillar structures of poly(lactic acid) and poly(vinyl alcohol) blend under various spinning conditions (*in preparation*)

Oral Presentations

1. N.H.A. Tran, H. Brünig, C. Hinüber, G. Heinrich. Melt spinning of nano fibrillary structures from immiscible poly (lactic acid) and poly (vinyl alcohol) blends. The 29th International Conference of the Polymer Processing Society, July 15-19, **2013** Nürnberg, Germany.
2. N.H.A. Tran, H. Brünig, R. Boldt, G. Heinrich. Morphology development of poly (lactic acid) and poly (vinyl alcohol) blend filaments along the spinline in melt spinning process. The 30th international conference of the Polymer Processing Society, June 8-12, **2014** Cleveland, Ohio, USA.

List of publications

3. N.H.A. Tran, H. Brünig, R. Boldt, G. Heinrich. Melt Spinning of Micro- or Nano-fibrillar Composites (MFCs) Based on Biodegradable Polymer Blends. The 1st Joint Turkey-Germany Workshop on Polymeric Nanocomposites, August 29-31, **2014** Istanbul, Turkey.
4. N.H.A. Tran, H. Brünig, M. Auf der Landwehr, G. Heinrich. Controlling the micro-and nanofibrillar morphology poly (lactic acid) and poly (vinyl alcohol) blend under various spinning conditions. The Polymer Processing Society Conference 2015, September 21-25, **2015**, Graz, Austria.

Poster Presentations

1. N.H.A. Tran, H. Brünig, C. Hinüber, G. Heinrich, Melt spinning of nano fibrillary structures based on biodegradable polymer blends. The 6th Aachen-Dresden International Textile Conference, Nov. 29-30, **2012** Dresden, Germany.
2. N.H.A. Tran, H. Brünig, G. Heinrich, Monitoring morphological changes of polymer blends along the spinnline using fiber capturing device, the 8th Aachen-Dresden International Textile Conference, Nov. 27-28, **2014** Dresden, Germany.

Other References

1. Schmelzspinnen von nanofibrillären Fasern auf der Basis von biologisch abbaubaren Polymer Blends. Presseinformationen, p.11, the 6th Aachen-Dresden International Textile Conference, Nov. 29-30, **2012** Dresden, Germany
2. N.H.A. Tran, H. Brünig, Erzeugung mikro- und nanofibrillärer Strukturen beim Schmelzspinnen, Jahresbericht **2013** des Leibniz-Instituts für Polymerforschung Dresden e.V., p. 65.
3. IPF intern, Heft 62, Januar **2013**, p. 5: Schmelzspinnen von nanofibrillären Fasern aus biologisch abbaubaren Polymer Blends.
4. Morphologieentwicklung beim Schmelzspinnen von Polymerblends sichtbar gemacht. Presseinformationen, p.10, the 8th Aachen-Dresden International Textile Conference, Nov. 27-28, **2014** Dresden, Germany.

Curriculum Vitae

Trần Nguyễn Hoài An
Born on August 20th 1980
Đồng Nai, Việt Nam
Family status: Married, two daughters

Education

10. 2011 — 03.2016   Ph.D. study at Technische Universität Dresden (TUD)/Leibniz-Institut für Polymerforschung Dresden e. V. (IPF Dresden e. V.)
Thesis title: Melt spinning and chacracterization of biodegradable nanofibrillar structures from poly(lactic acid) and poly(vinyl alcohol) blends
Supervisors: Prof. Gert Heinrich, Dr. Harald Brünig

10.2005 — 04.2008   Master study of textile and clothing technology at TUD (2 year courses)
Degree: M.Sc. in textile and clothing technology
Thesis title: Investigating the effect of polymer structure of man-made fibre yarns on the occurrence of the periodical structural variations of circular knitted fabrics
Supervisors: Prof. Chokri Cherif, Dr. Rolf-Dieter Hund, Dipl. André Matthes

1998 — 2003   Bachelor study of mechanical engineering (4.5 year courses) (major: textile and clothing technology)
Degree: B.-Eng. in mechanical engineering

Work experience

2. 2012 — now   Research associate at IPF Dresden e. V., Institute of polymer materials (Prof. Gert Heinrich), Department of processing (Prof. Udo Wagenknecht), Research group of fiber spinning (Dr. Harald Brünig)
DFG Project: „Entwicklung eines neuartigen Filamentgarnes" (BR 1886/6-1)

2008 — 2011   Lecturer at Ho Chi Minh City University of Technology (HCMUT), Faculty of mechanical engineering (FME), Department of textile and clothing engineering (DTCE), Ho Chi Minh City (HCMC), Viet Nam

2003 — 2005   Assistant lecturer at HCMUT, FME, DTCE, HCMC, Viet Nam

Awards

2011 — 2015   Ph.D. Scholarship of the Ministry of Education and Training of Viet Nam
2005 — 2008   M.Sc. Scholarship of Katholischer Akademischer Ausländer-Dienst (KAAD)
4.2003   Silver medal for excellent student - studying period 1998-2003 honor, granted by the rector of HCMUT, HCMC, Viet Nam